IMAGES
of America

OMAHA'S TRANS-MISSISSIPPI EXPOSITION

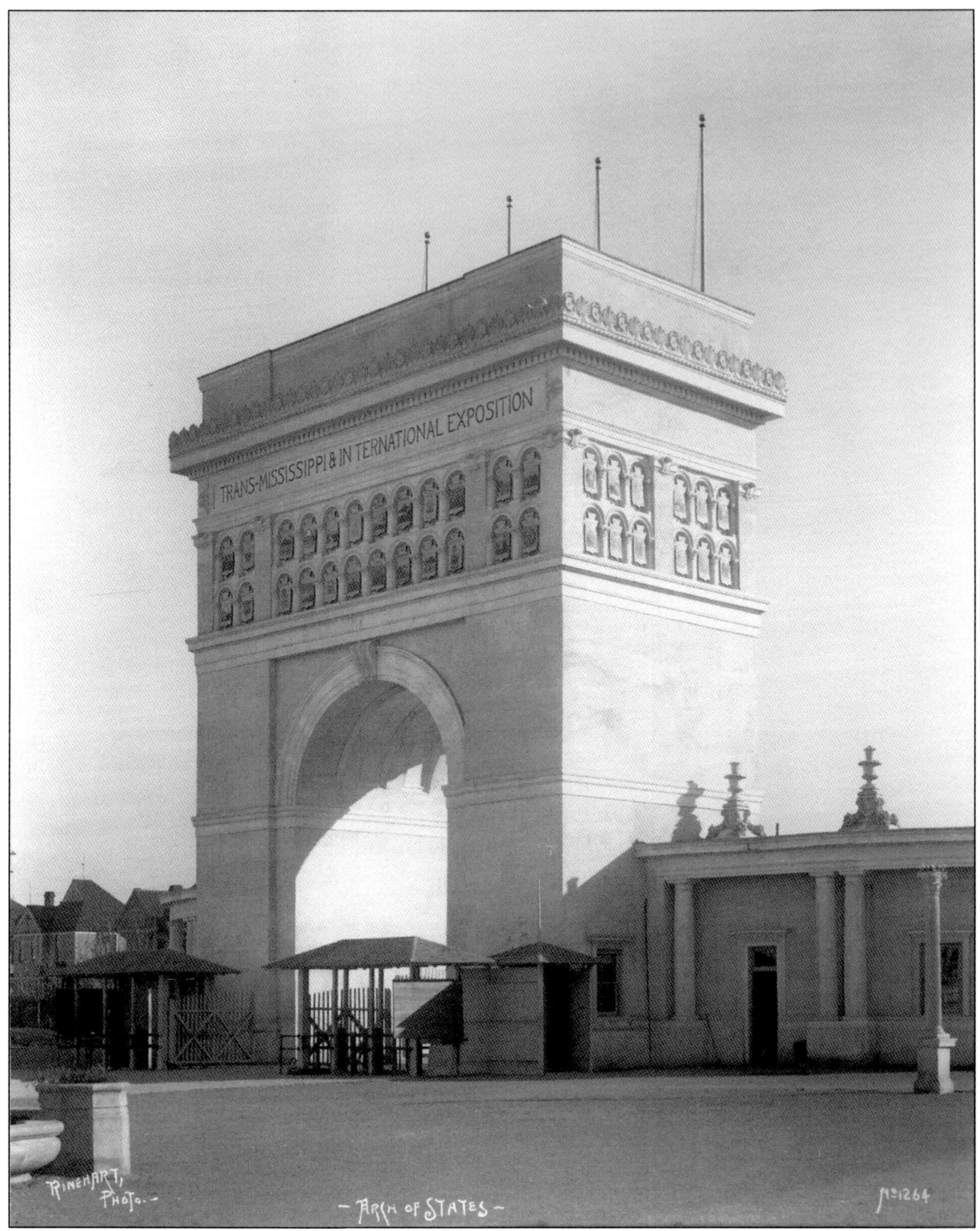

Emblazoned across the top of the Arch of the States were the words "Trans-Mississippi & International Exposition." The coats of arms of the states located in this region surrounded the upper section of the arch. The arch also served as the main ticket office of the exposition and one of seven entrances. Single-day tickets were 25¢ for children ages 5 to 12, and 50¢ for adults. (Photo courtesy of Omaha Public Library.)

IMAGES
of America

OMAHA'S TRANS-MISSISSIPPI EXPOSITION

Jess R. Peterson

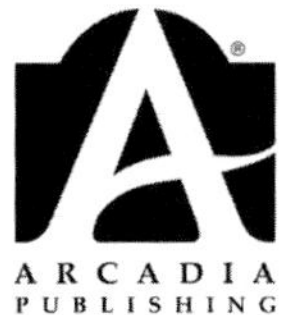

ISBN 978-0-7385-3151-9

Published by Arcadia Publishing
Charleston, South Carolina

Printed in the United States of America

Library of Congress Catalog Card Number: 2003106976

For all general information contact Arcadia Publishing at:
Telephone 843-853-2070
Fax 843-853-0044
E-mail sales@arcadiapublishing.com
For customer service and orders:
Toll-Free 1-888-313-2665

Visit us on the Internet at www.arcadiapublishing.com

For Mom, for all you have done.

A legacy fulfilled, for Dad.

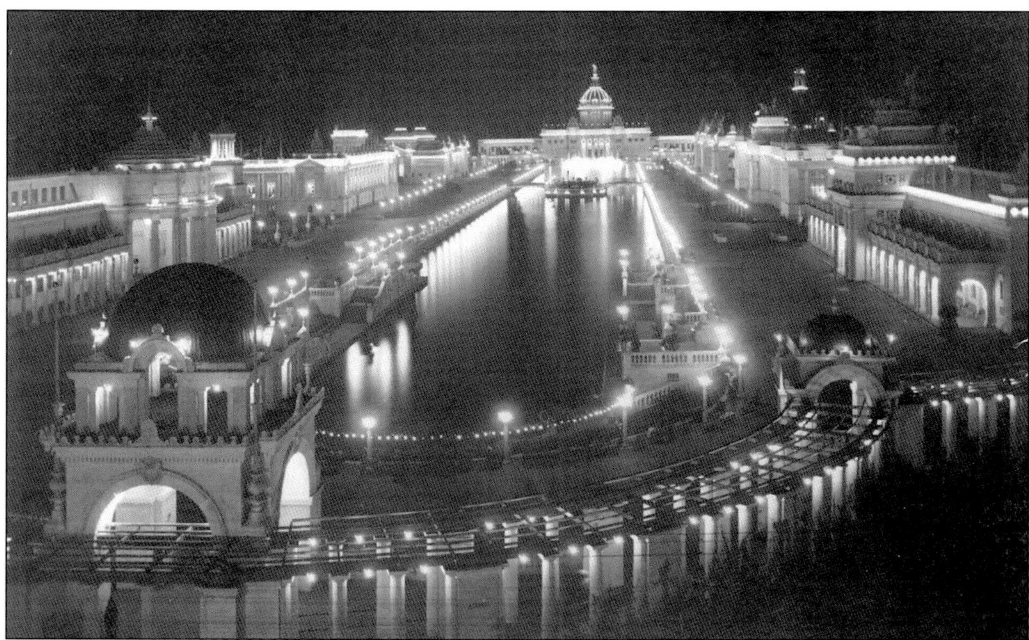

Illumination experts used more than 20,000 light bulbs to complete their work of lighting the Grand Court. Rooftops and other exposition structures were carefully outlined with 8- and 16-candle power bulbs. James B. Haynes, author of the official exposition history, described the nighttime illumination of the Grand Court as moonlight without shadows. (Photo courtesy of Omaha Public Library.)

Contents

Acknowledgments

Just as the Trans-Mississippi Exposition was a community effort, it is with considerable pride that I acknowledge the contributions others have made to this work.

Thank you to the Omaha Public Library for opening their photographic resources in support of this effort. The effort of library staff members to preserve the official exposition photographs of Frank A. Rinehart is an example of historic preservation worthy of the highest honor. Thank you to Tom Heenan and Joanne Ferguson-Cavanaugh at the library for their individual efforts to secure many images within this work.

Thank you to Frank A. Rinehart for sharing a photographic vision of our shared history. Thank you to James B. Haynes and John Wakefield for recording the history of this grand event. The writings of these gentlemen are invaluable resources for continued study of the exposition.

On June 14, 1898, Gurdon W. Wattles, President of the Trans-Mississippi and International Exposition, spoke at the dedication ceremony for the exposition building representing the citizens of Nebraska. Wattles said, "To the state of Nebraska, the future historian will give the credit of erecting, in times of adversity, a great exposition, destined to break down prejudices, build up commerce, and promote peace and good will throughout the land." It is in the spirit of this hope that this work is presented to the historians of a different era and the optimists of a different generation, many years removed from their counterparts, who shared a common dream.

Jess

INTRODUCTION

In an era of fairs and grand expositions, the Trans-Mississippi and International Exposition provided Omaha, Nebraska, with an opportunity to represent the economic, scientific, and cultural achievements of nineteenth-century America. Staged on the heels of a national economic crisis, a drought that severely impacted farm production, and during a time of national military conflict, the 1898 Trans-Mississippi Exposition remains an event by which all others will be measured as the history of Omaha continues to be written.

The explosion of the *U.S.S. Maine*, the spark that ignited the Spanish-American War, occurred less than four months prior to the opening of the Trans-Mississippi Exposition. While exposition organizers debated the wisdom of a grand celebration during wartime, public subscriptions and donations in excess of one-half million dollars (of which 90 percent was returned to shareholders at the conclusion of the exposition) signified a high degree of support.

The developing art of photography found a stunning subject in the spectacle of the Trans-Mississippi Exposition. Preserved by the Omaha Public Library, the images of official exposition photographer Frank A. Rinehart are a priceless visual reminder of this event. Represented in these photographs are not only the successes of those individuals who pioneered the American West, but their descendants, who dashed former notions of the West as the "Great American Desert."

The exposition, comprised of approximately 110 temporary structures, large and small, was one of displays, thoughts, and ideals, as well as entertainment. The Grand Court of the exposition featured the most significant structures. These grand edifices, reminiscent of castles and other imposing architectural monuments throughout the world, welcomed visitors to study the achievements of a nation at the crossroads of an era that would soon yield to the advances of a modern age. Included among over 5,000 exhibits were works of fine art, household products, modern tools, industrial equipment, natural resources, a diverse array of agricultural products, and the institutions of government.

Three areas provided exposition visitors with a variety of exhibits and opportunities for entertainment. The Kountze tract, named for Herman Kountze, a key Trans-Mississippi benefactor, was the centerpiece of the exposition. This area featured the magnificence of the Grand Court and the largest proportion of significant exhibits. The Bluffs tract, to the immediate east of the Grand Court, included the Horticulture Building, nine state buildings, the East Midway, and the showplace of the Trans-Mississippi Exposition, the Grand Plaza. Represented within the walls of the state buildings were the achievements and traditions of states that sought greater representation than could be provided within the confines of other exposition buildings. The North tract included additional midway attractions, the Transportation Building, and the area of the Indian Congress held in conjunction with the exposition.

Before the era of flight, perhaps the finest view of the exposition was available from atop a popular midway attraction, the Giant See-Saw. From this vantage point, riders looked across an exposition that, four years earlier, might have been regarded as a distant dream for the citizens of Omaha, Nebraska. Recognizing the inevitability of comparison with other events—particularly the successful 1893 World's Columbian Exposition, held in Chicago, Illinois—Trans-Mississippi organizers sought to present the American West as an area of vast resources, economic growth, and virtually unbounded development potential. What began as a showcase of the West, however, soon became a national and international celebration of the highest degree. Presented on June 1, 1898, during the opening day ceremonies of the Trans-Mississippi Exposition, the Song of Welcome and Ode to the Exposition included the following verse:

Welcome, thrice welcome, to the people of our land;
Welcome to the people, the people of the world;
Here, North and South and East and West, united hand in hand,
Have reared a city and their flag unfurled.
Welcome, welcome, welcome, to the people of the world!

The need for diversion from the thought-provoking aspects of this event was not overlooked. The midway featured thrill rides, performers, cultural and historical exhibits, and a choice of restaurants for visitors. In an effort to preserve the high ideals of the Trans-Mississippi Exposition, organizers sought to present midway attractions that would stimulate and entertain visitors, without succumbing to interests in support of including attractions that might be regarded as mere carnival sideshows.

Photographs of Native Americans in attendance at the Indian Congress, which was held during the last half of the exposition season, are a cultural historic record that rises in significance with the passage of time. Members of numerous tribes arrived on the exposition grounds under an air of anticipation like none other. The curiosity of exposition visitors was soon replaced by fascination. For many thoughtful visitors, the residual effects of first-hand contact with this unfamiliar culture dimmed the bright glow of stereotype.

To measure the success of such an undertaking is, in many respects, more monumental than the combined tasks of envisioning, constructing, and promoting the exposition itself. Success may be measured in terms of the 2.6 million visitors who spent at least part of their visit entranced in study of the exhibits. Each visitor, including the great-grandparents of the author of this volume, left the exposition with a story to be shared with future generations. Proximity to peers created economic ties that bound previously unfamiliar interests for many years to come.

Success can also be measured in more apparent terms. The United States Government Building stood as a testament to the national importance of the Trans-Mississippi Exposition. A series of commemorative stamps issued by the United States Postal Service served as a messenger of the ideals put forth by the exposition. The financial success of the exposition, far from certain at its inception, remains a source of pride for Omaha.

On the closing day of the exposition, October 31, 1898, designated as Omaha Day, Mayor Frank E. Moores challenged Omahans to "look forward to the future with faith and courage and let us one and all put our shoulders to the wheel of Omaha prosperity and progress." Somewhere within the crowd on that day and many others were children who, inspired by what they saw, developed the faith and courage to fulfill this wish.

One

GLORIA ACTIONES CINXIT

On September 10, 1902, medals were awarded to President Gurdon Wattles and the remaining six members of the Trans-Mississippi and International Exposition Board of Directors. The figure of Victory depicted on the medal held a scroll featuring the phrase "Gloria Actiones Cinxit." Latin for "Glory surrounds his conduct," this phrase represented the pride with which members of the Board recalled, on this night, the actions of all who had participated in their undertaking.

The grounds of the Trans-Mississippi Exposition encompassed 184 acres of land within the city of Omaha. A complete tour of the exposition was clearly a multiple-day prospect for the average visitor. Entering the grounds of the exposition at its main entrance, the Arch of the States, visitors were immediately greeted with the visual treat of the Grand Court. The architectural feats of this area provided a worthy stage for the displays inside of the structures. A walk around the Grand Court took visitors on an estimated one-mile tour through some of the most spectacular exposition sites.

Additional exposition attractions included daily performances, speeches, and music played in the Auditorium, as well as the outdoor Grand Plaza. Not to be forgotten was the midway, which included attractions to delight, entertain, and amaze all who strolled its streets. Finally, no day at the exposition was complete without remaining for the nighttime events. Frequent displays of fireworks and the spectacular lighting of the Grand Court were sights no visitor would soon forget.

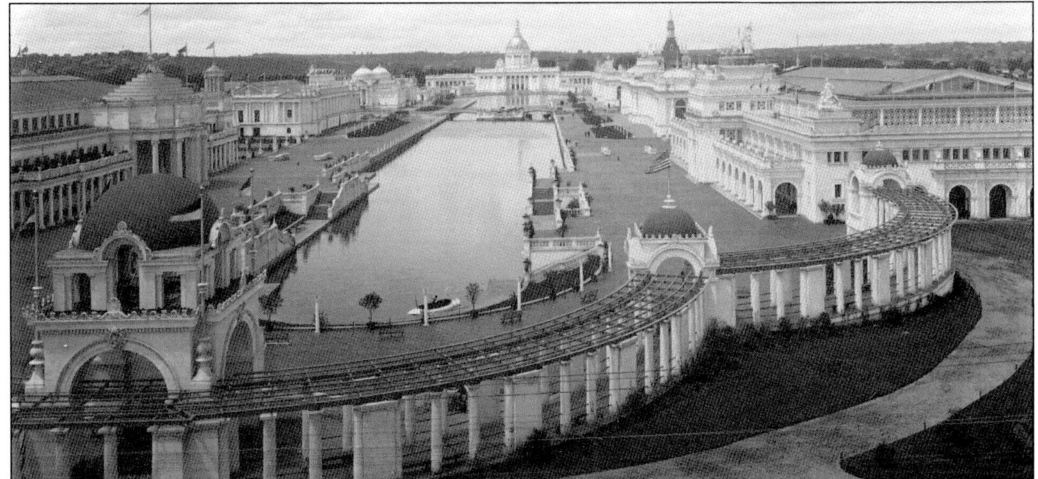

The Kountze tract, named for exposition treasurer Herman Kountze, served as the location for the Grand Court of the Trans-Mississippi Exposition. Approximately 2,590 feet long and 680 feet wide, the Grand Court was bound by Sherman Avenue, Twenty-Fourth Street, Pinkney Street, and Pratt Street on the east, west, south and north sides, respectively. While housing has since filled in much of the former exposition grounds, Omaha's Kountze Park, constructed within these confines, is the last open space remaining from the Trans-Mississippi Exposition. (Photo courtesy of Omaha Public Library.)

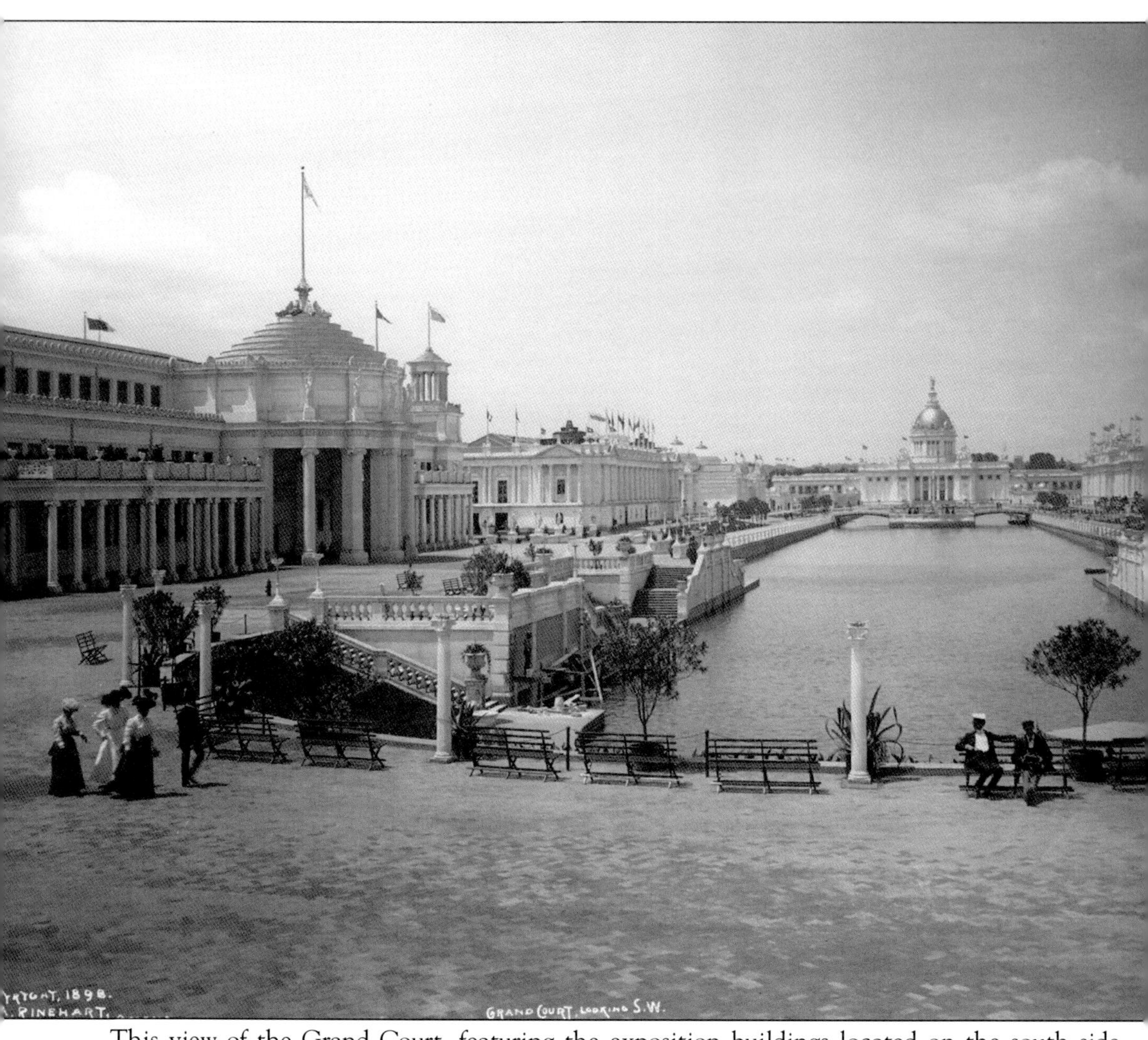

This view of the Grand Court, featuring the exposition buildings located on the south side of the lagoon, illustrates the efforts of exposition architects, who constructed nearly a mile of sheltered walkway, combining buildings with a series of colonnades and interconnecting courtyards. Entering the exposition grounds, some visitors, clearly in awe of the sights, reported feeling as if they had been transported to Venice. (Photo courtesy of Omaha Public Library.)

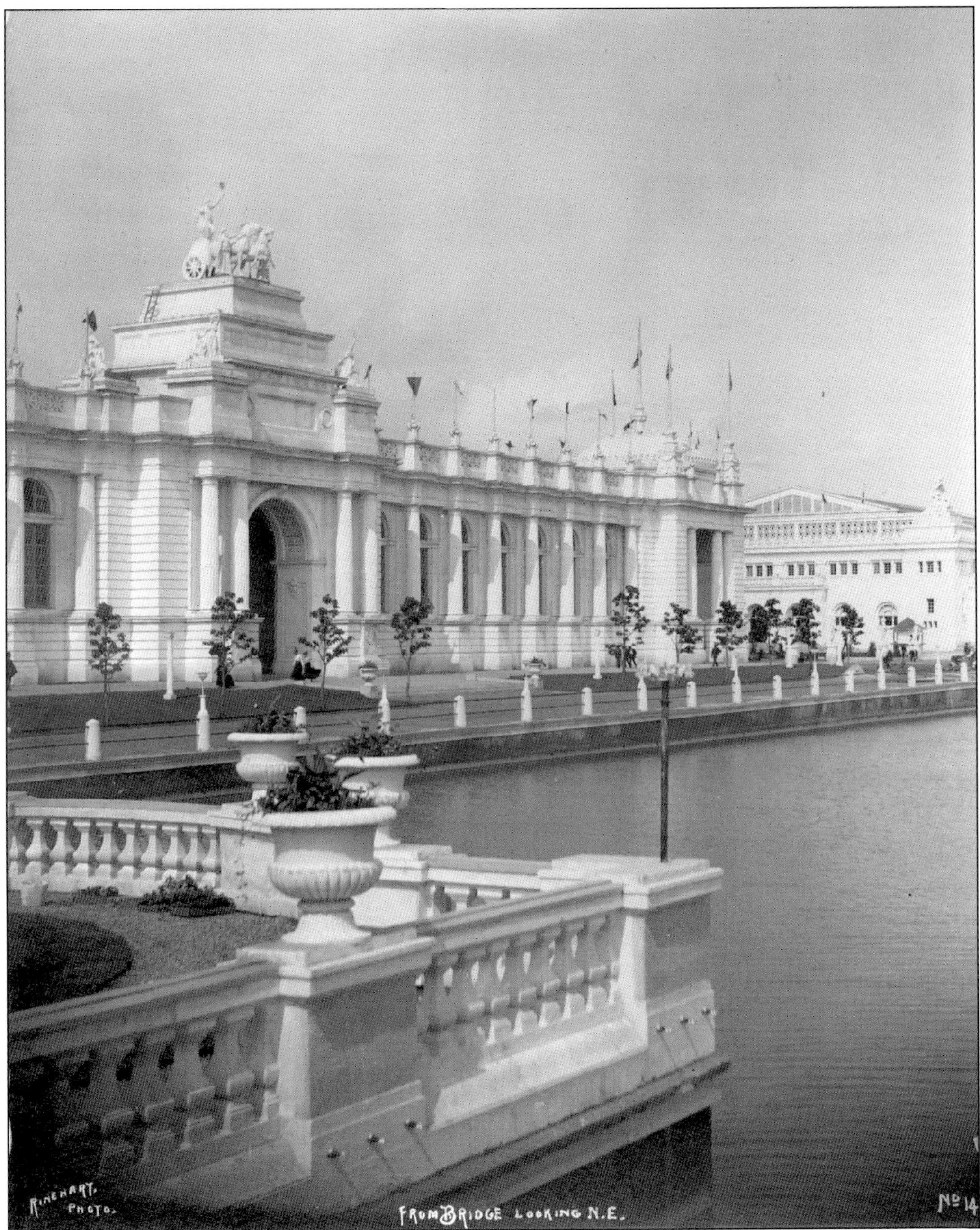

This view from the bridge at the center of the lagoon shows the landscaping features in place in front of the Manufactures Building. Approximately 1,200 shade trees lined the walks of the Trans-Mississippi Exposition. Nine thousand feet of sod, along with a variety of other plants and flowers, provided a natural complement to the ivory facades of exposition buildings. (Photo courtesy of Omaha Public Library.)

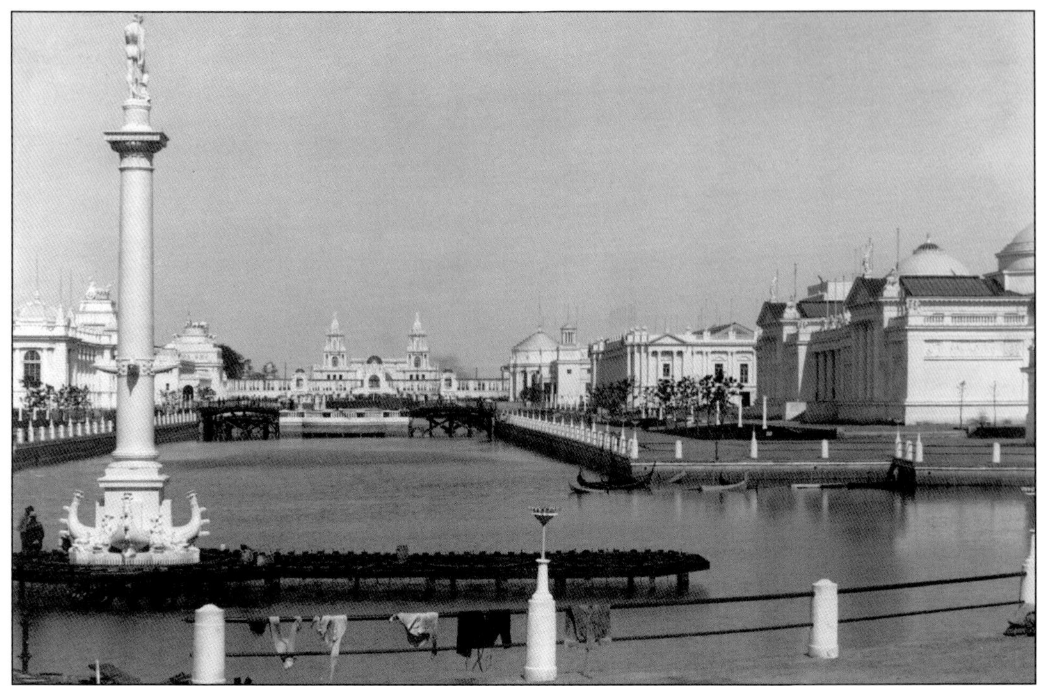

Looking east from the court in front of the United States Government Building, located at the west end of the Grand Court, provided visitors with this spectacular architectural overview of the area. The focal point of the mirror basin was the "Nautilus" fountain, featuring a likeness of the Roman god of the sea, Neptune, atop a single pillar. Colored lights created a magical show for those who remained on the exposition grounds during the evening hours. (Photo courtesy of Omaha Public Library.)

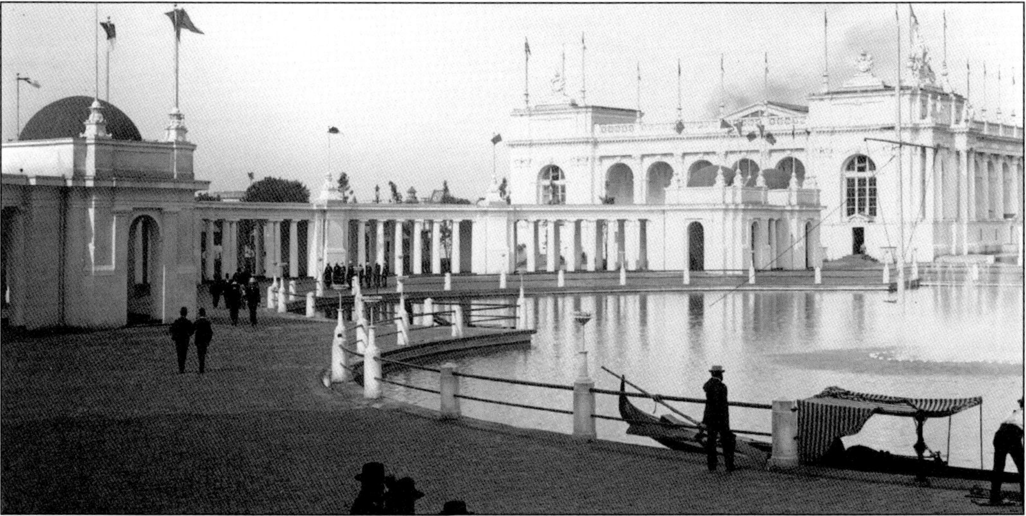

The colonnade connecting the United States Government Building with the Agriculture Building, pictured at right, was a key feature in the Classical design of exposition structures. Reminiscent of St. Peter's Square in Rome, the colonnades provided covered access to the government building. The bank of the mirror basin also served as one docking point for the scenic gondola ride on the lagoon. (Photo courtesy of Omaha Public Library.)

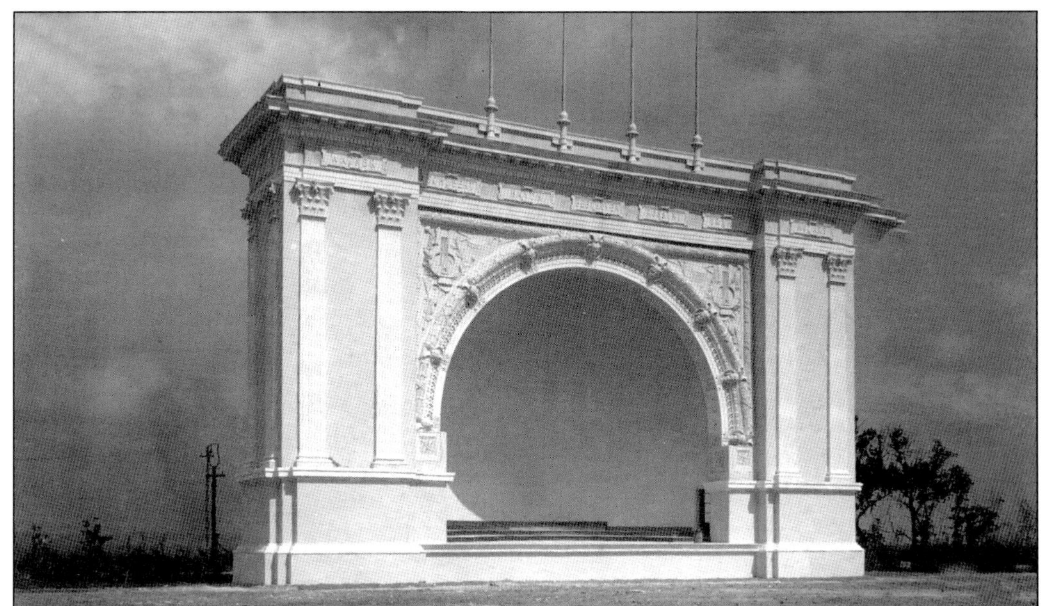

The Music Pavilion, located immediately east of the Grand Court in the exposition area known as the Grand Plaza, was the site of afternoon and evening concerts. Among many local and regional favorites, the United States Marine Band was a favorite of exposition visitors. The daily concerts had the collateral benefit of bringing together exposition visitors, who initiated lasting relationships to advance the understanding and cooperation between those living inside and outside of the Trans-Mississippi region. (Photo courtesy of Omaha Public Library.)

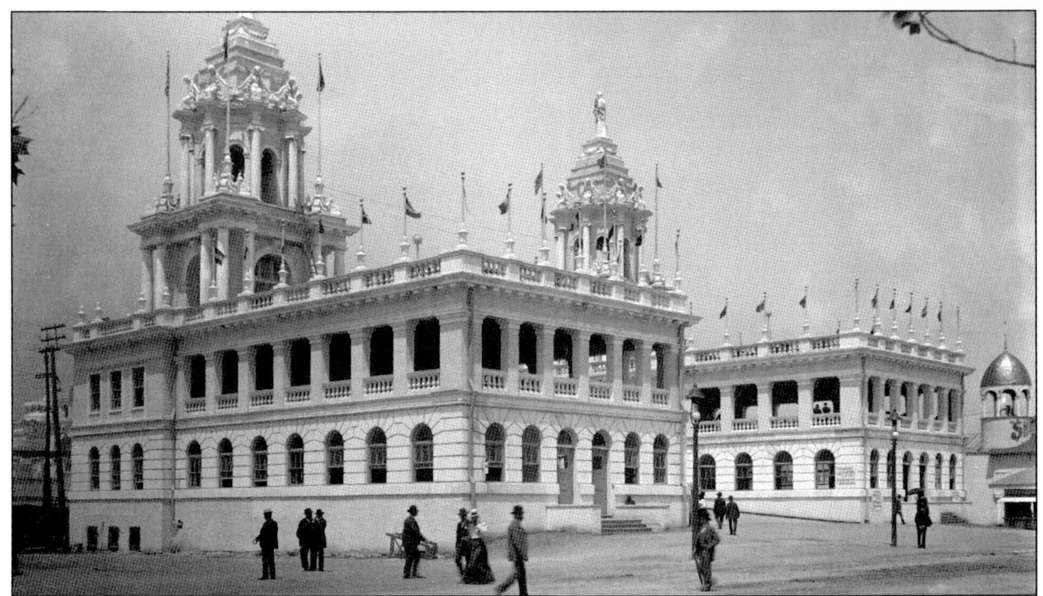

The twin restaurants, featuring matching terrace towers, offered food service on three levels at the South Viaduct, which crossed Sherman Avenue at the east end of the Grand Court. The South Viaduct, pictured between the two prominent structures, served as a primary route between the main exhibits of the exposition, the state buildings of the Bluffs tract, and the entertainment of the east midway. (Photo courtesy of Omaha Public Library.)

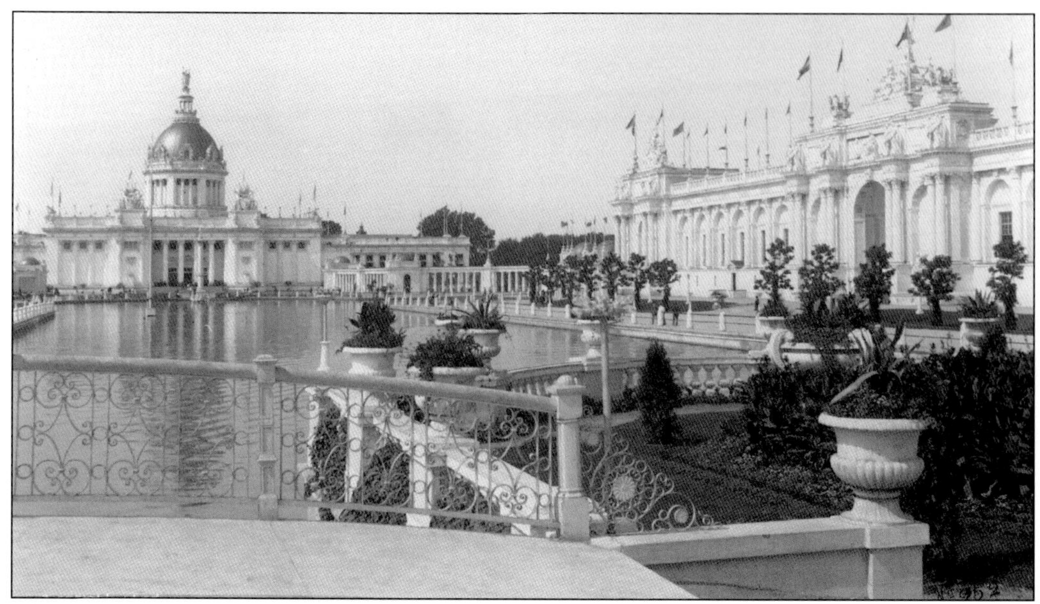

The many fine details of the exposition grounds were visible from this vantage point on the bridge providing passage at the center of the lagoon. The United States Government Building, pictured at the left, was often compared to both the U.S. Capitol Building and St. Peter's Basilica. Also visible from this point was the Renaissance-influenced design of the Agriculture Building. (Photo courtesy of Omaha Public Library.)

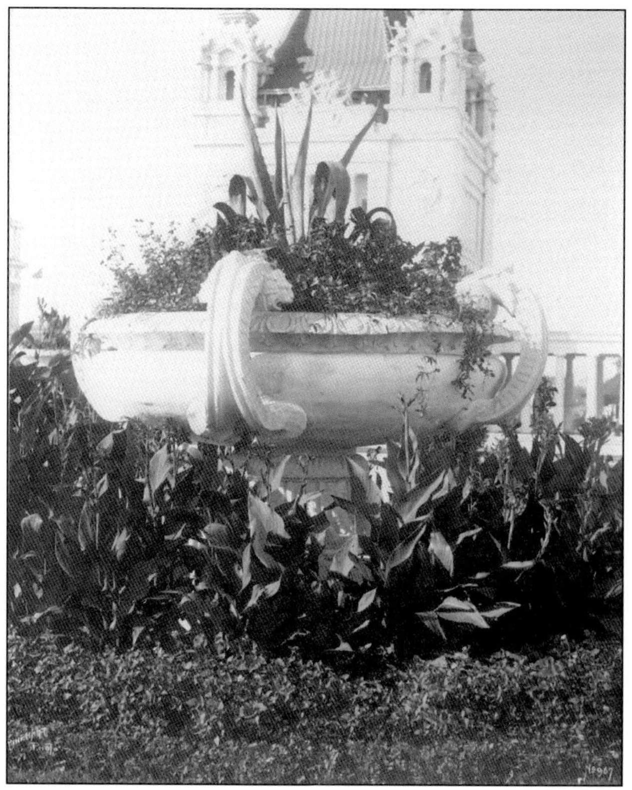

Matching the ivory tones of exposition buildings were the ubiquitous flower vases. Eighty-five such vases graced the banks of the exposition lagoon, as well as building entrances. A greenhouse, constructed for the specific purpose of raising approximately 100,000 flowers and plants to be displayed at the exposition, was the site of much activity during the winter of 1897–1898. (Photo courtesy of Omaha Public Library.)

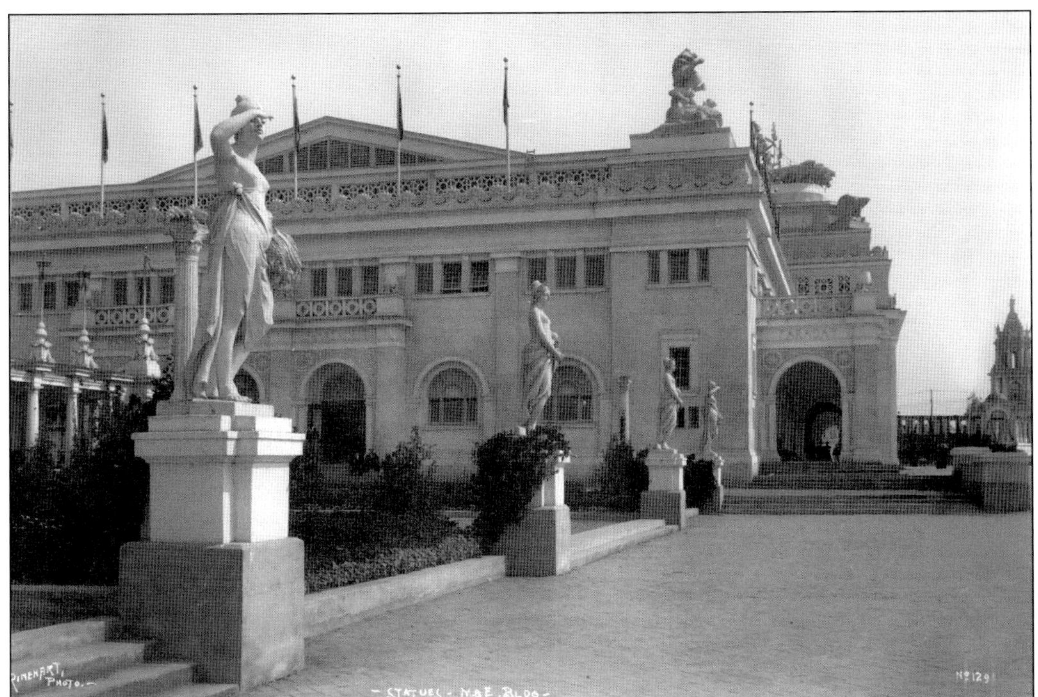

This line of statues and Pompeian colonnade, featured in the courtyard joining the Machinery and Electricity Building with the Manufactures Building, provided a relaxing venue for exposition visitors, as well as adding beauty and a sense of style to the exposition grounds. (Photo courtesy of Omaha Public Library.)

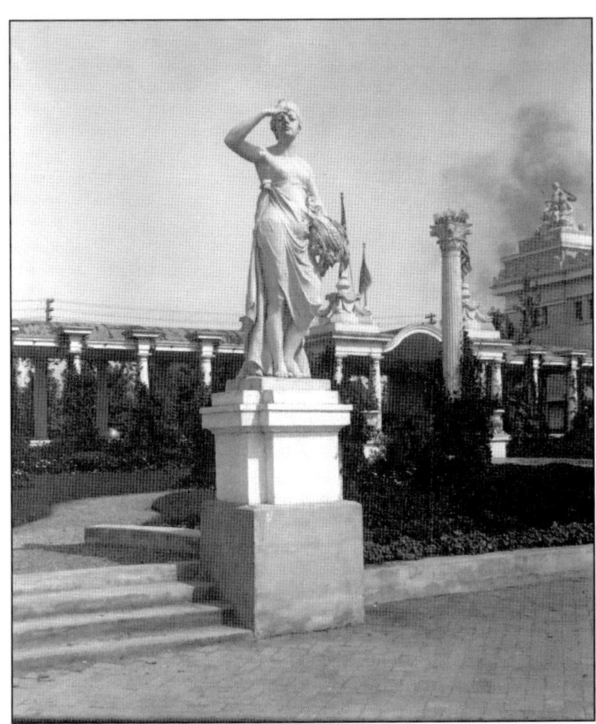

This statue of Ceres, the Roman goddess of agriculture, stood near the Machinery and Electricity Building. Statuary elements such as this were important features of the gardens and courtyards joining the most significant exposition buildings. Pictured in front of the statue is a section of the brick walking surface connecting all points of the Grand Court. (Photo courtesy of Omaha Public Library.)

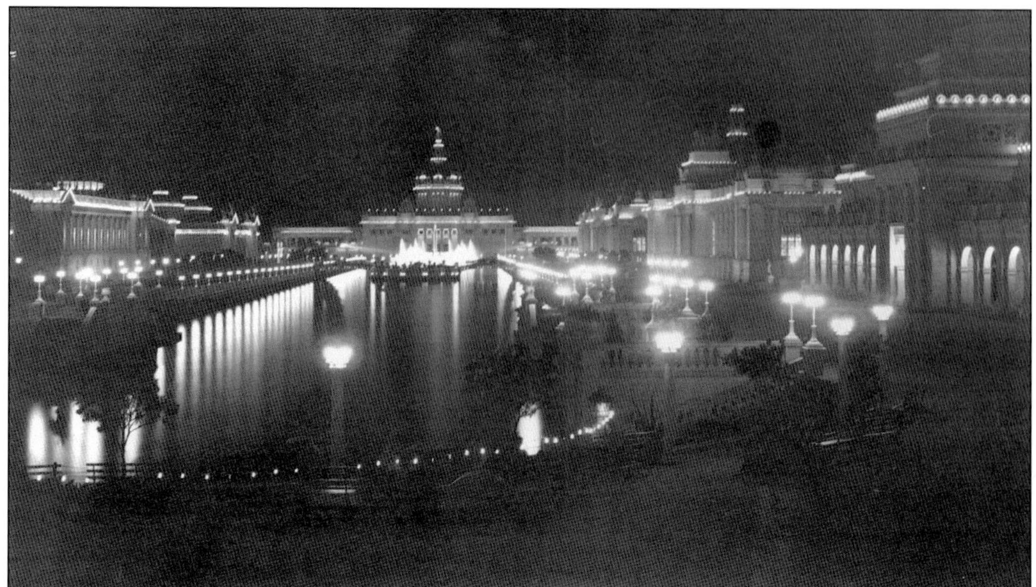

Upon the inspiration of Luther Stieringer, Omahan Henry Rustin achieved what is arguably the most spectacular feature of the Trans-Mississippi Exposition: the nighttime illumination of the Grand Court. Efforts to conceal light bulbs and an appreciation for the aesthetic beauty of reflected light were the guiding forces behind the fascination awaiting exposition visitors after dark. (Photo courtesy of Omaha Public Library.)

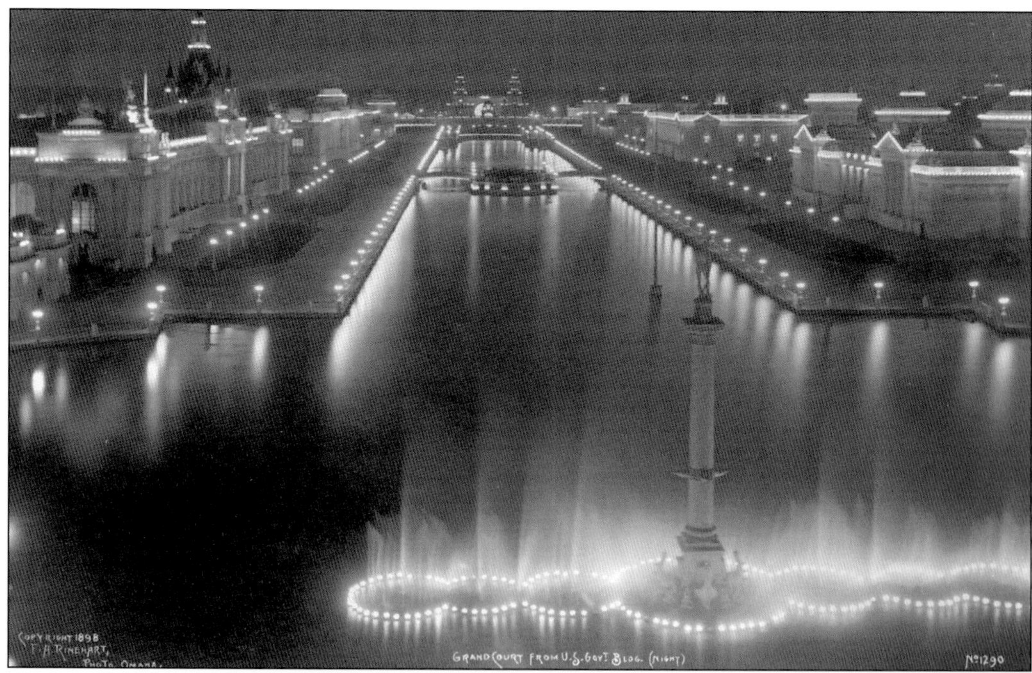

This spectacular nighttime view of the exposition, taken from atop the United States Government Building, must have been a scene that Frank A. Rinehart, official exposition photographer, hesitated to leave. Music, fireworks, and, of course, the illumination of the Court were among the many evening attractions for exposition visitors. (Photo courtesy of Omaha Public Library.)

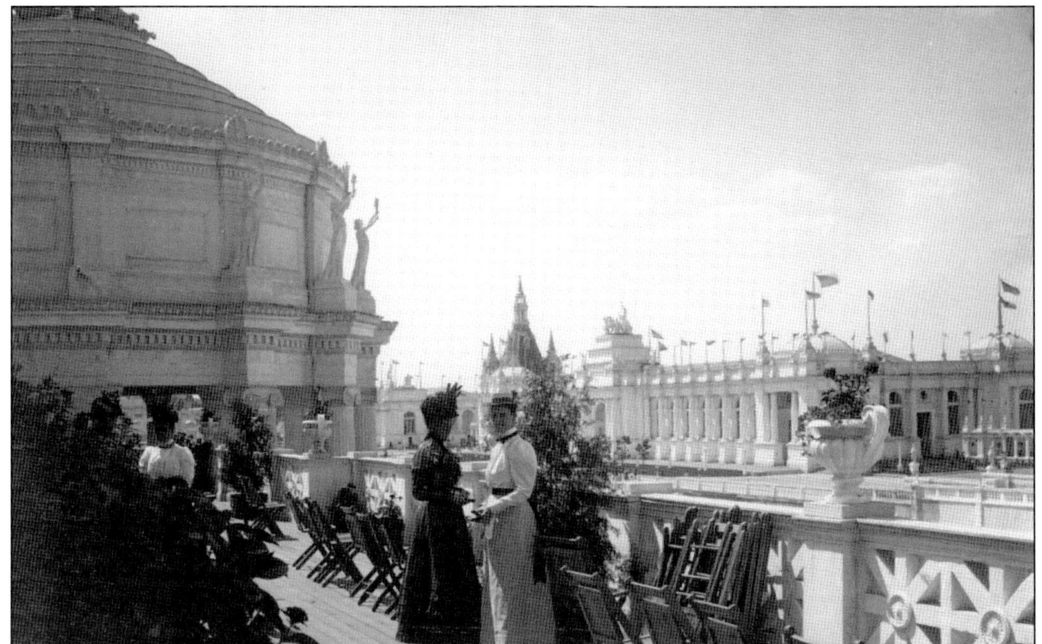

Exposition visitors could enjoy the view of the lagoon and the eastern half of the Grand Court from this balcony atop the Mines and Mining Building. The dome of this structure, pictured at the left of this photograph, rose to a height of 75 feet. During their visit, these ladies probably took the opportunity to enjoy the rooftop garden of the Mines and Mining Building. (Photo courtesy of Omaha Public Library.)

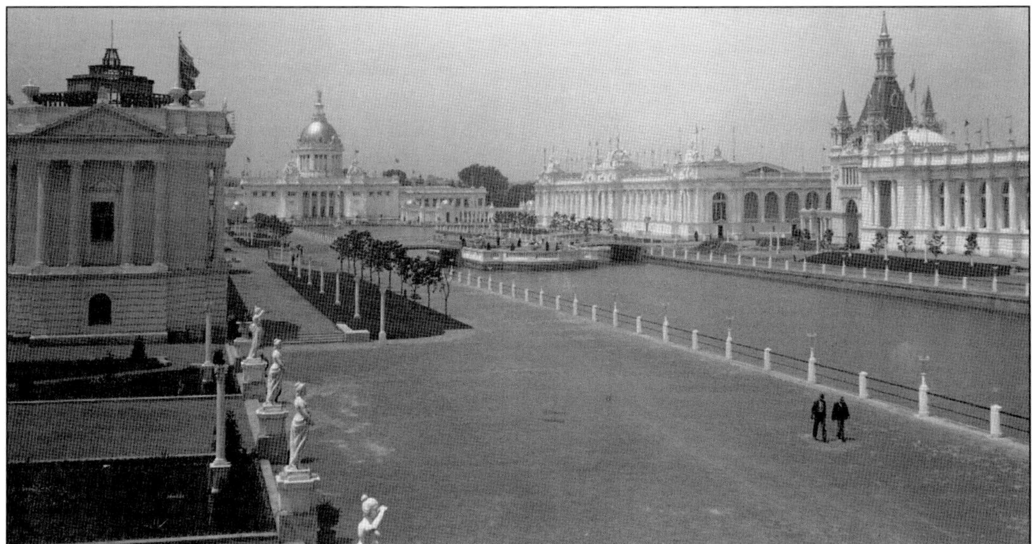

The central feature of the Grand Court was the lagoon. Measuring 2,000 feet in length and 150 feet in width, the lagoon held approximately seven million gallons of water. Managing the combined effects of Nebraska's summer rainstorms and periodic heat waves were a constant challenge for those charged with maintaining the lagoon. After initial attempts failed to supply the lagoon with water from an artesian well, the nearby Missouri River became an inexhaustible source of water. (Photo courtesy of Omaha Public Library.)

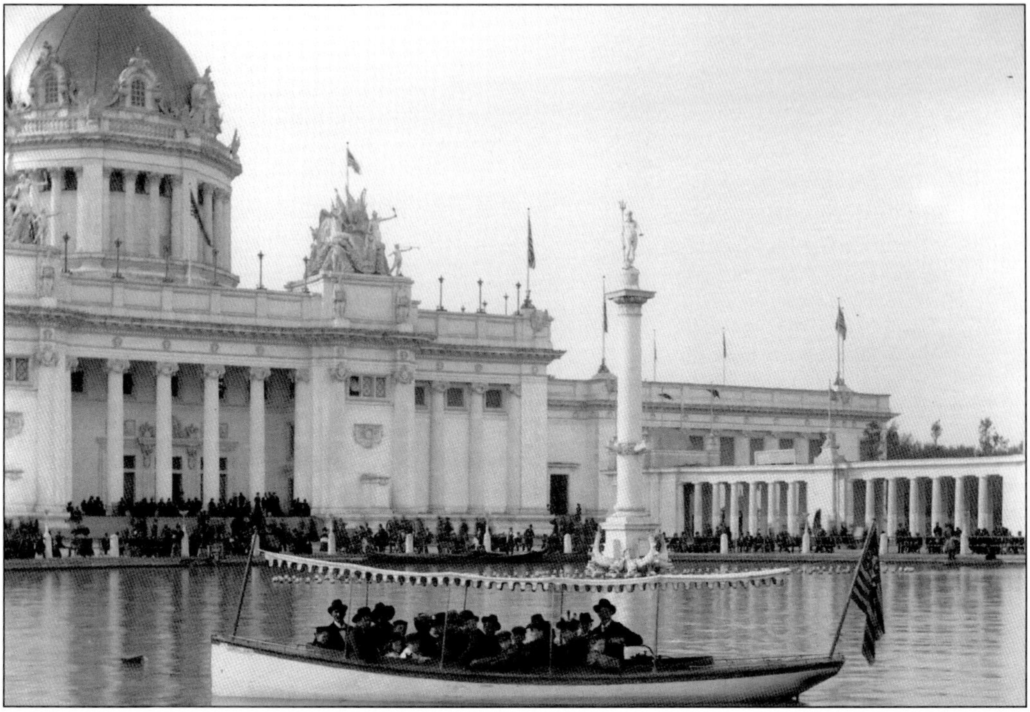

A leisurely gondola ride was one of the most enjoyable means for visitors to see the magnificence of the Trans-Mississippi Exposition. These visitors, at the center of the mirror basin, are facing the United States Government Building and the Nautilus statue. Riders were treated to the music and vocal talents of gondoliers during their tour. (Photo courtesy of Omaha Public Library.)

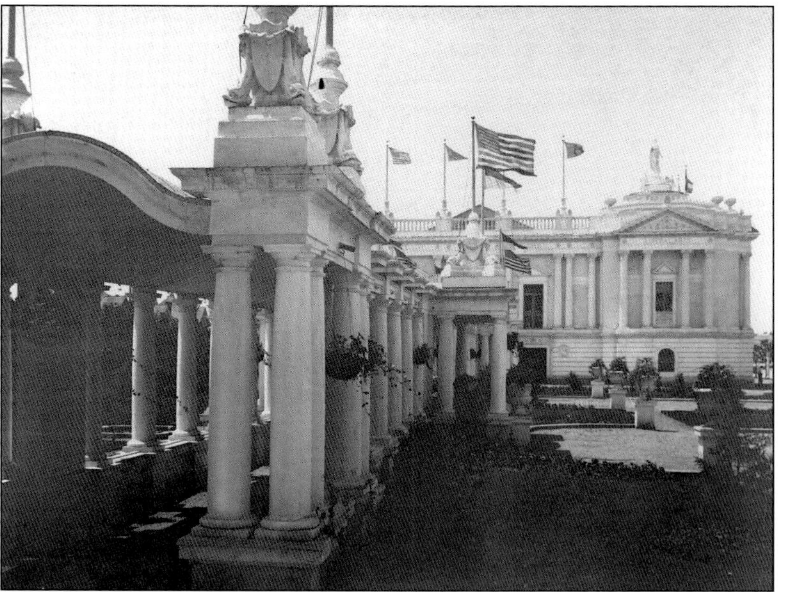

This colonnade served as a passageway between the Mines and Mining Building and the Liberal Arts Building (pictured). The colonnades were a key component used by exposition architects to meet their goal of focusing the attention of exposition visitors on the available exhibits. Visitors strolling the exposition grounds could become consumed in the sites, indoor and outdoor, while the achievements of the states comprising the Trans-Mississippi region were showcased in this regal setting. (Photo courtesy of Omaha Public Library.)

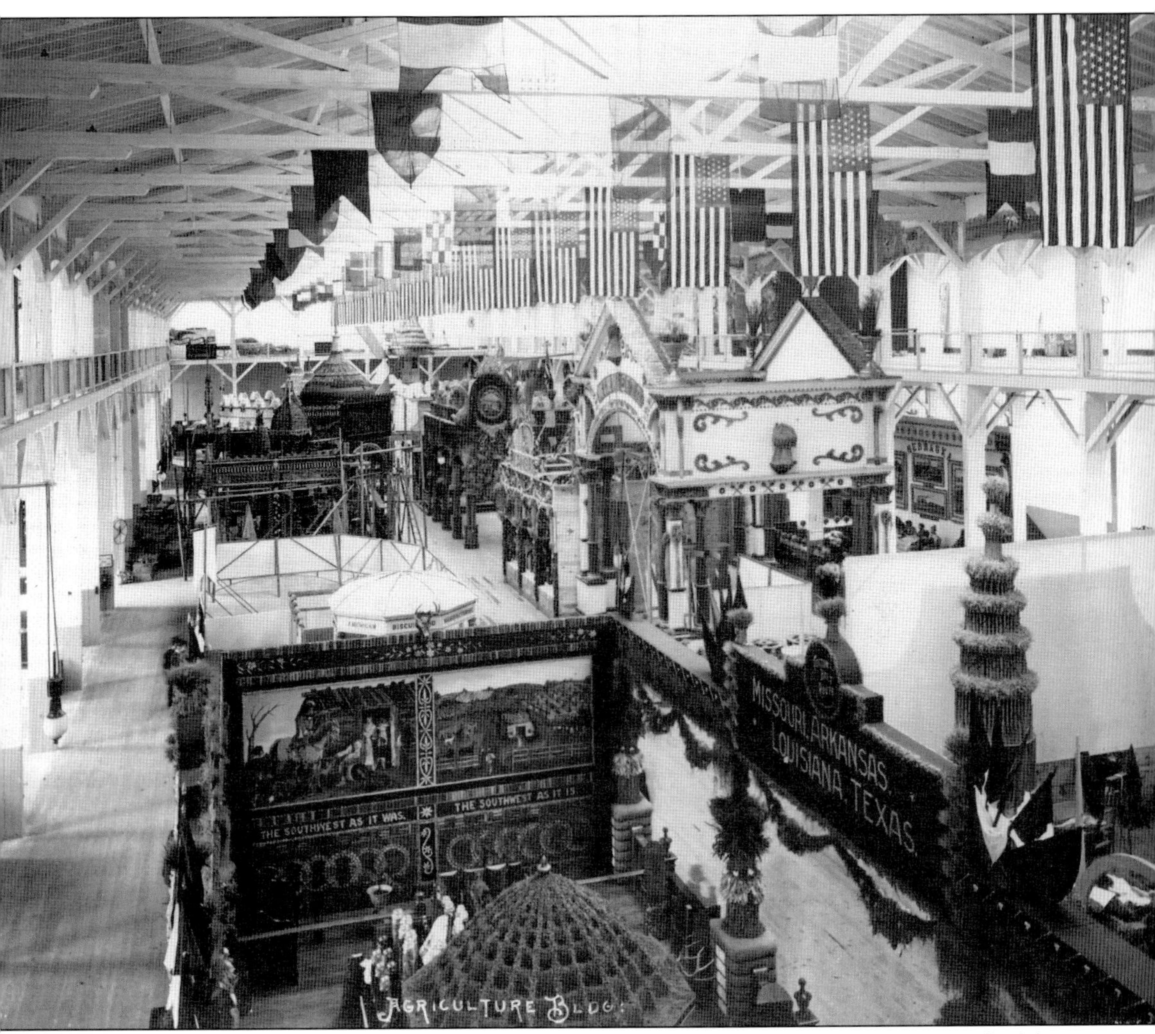

The agricultural products of the Trans-Mississippi region were featured in many colorful, creative displays, such as those seen in this view inside of the Agriculture Building. The exhibits of fifteen states and many commercial enterprises were featured in this structure. A friendly inter-state rivalry guided the often artistic display of farm products. (Photo courtesy of Omaha Public Library.)

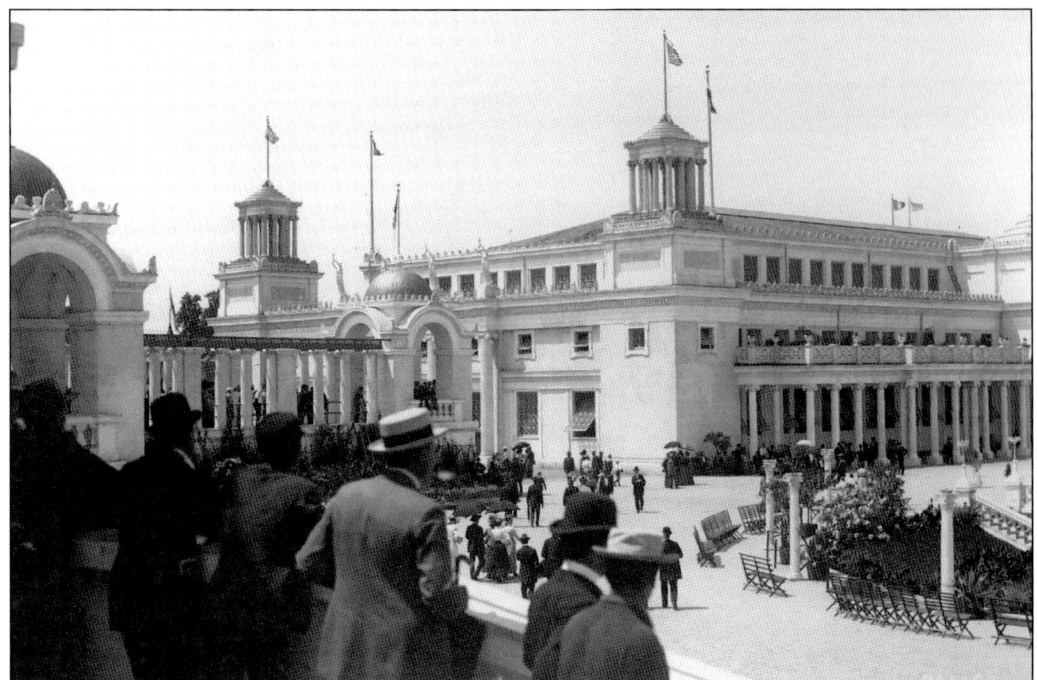

The opportunity to marvel at the spectacle of the exposition was, no doubt, the goal of these exposition visitors. Their vantage point atop the colonnade at the east end of the Grand Court provided this view of the Mines and Mining Building. Shade from the heat of a Nebraska summer was a welcome respite for these impeccably-dressed nineteenth century gentlemen. (Photo courtesy of Omaha Public Library.)

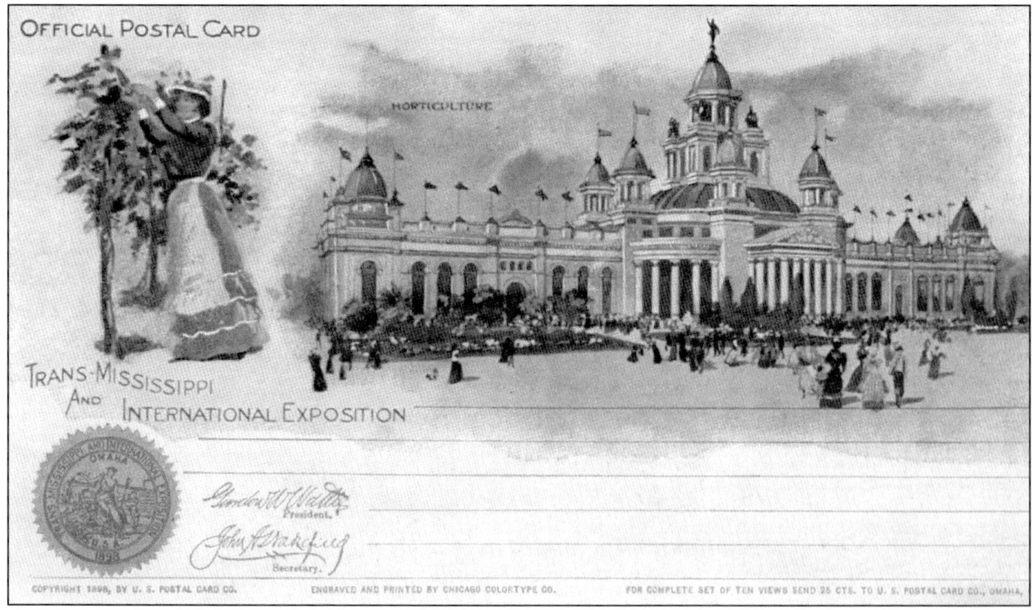

The Horticulture Building is featured on this postcard, one of ten printed by the U.S. Postal Card Company as the official collection of postcards from the Trans-Mississippi Exposition. (Photo courtesy of Omaha Public Library.)

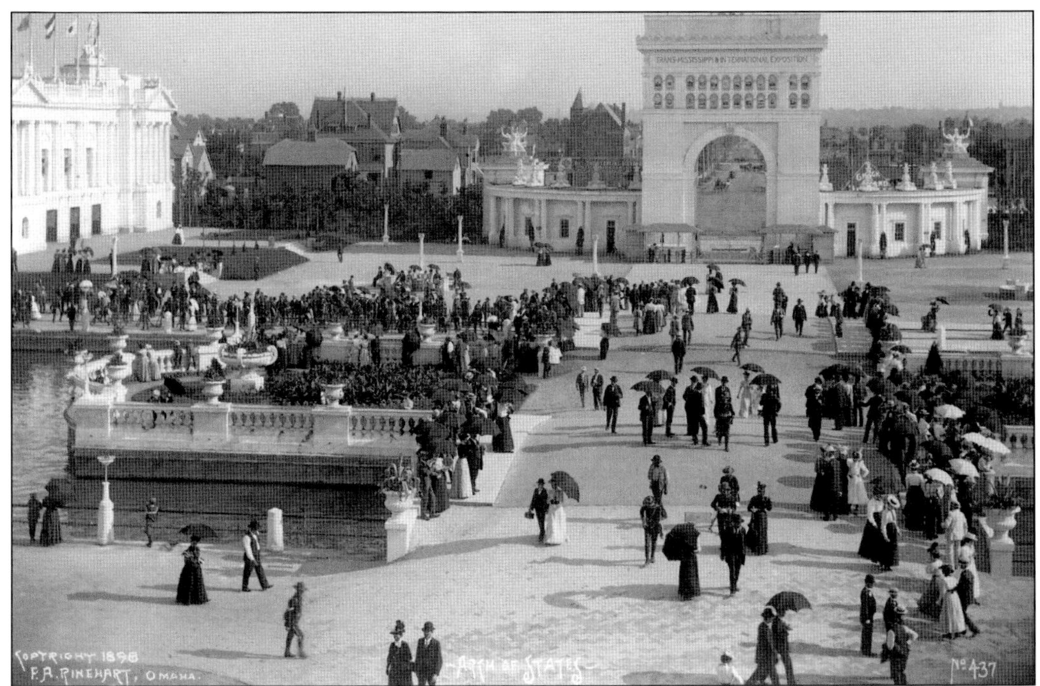

As the opening day parade approached the main entrance of the exposition at the Arch of the States, pictured here during the regular exposition season, the *Omaha Daily Bee* exclaimed, "Omaha's day of days has dawned." The two-mile long procession began at 10:30 a.m. in downtown Omaha at Sixteenth and Douglas Streets, and reached the Arch of the States at 11:00 a.m. (Photo courtesy of Omaha Public Library.)

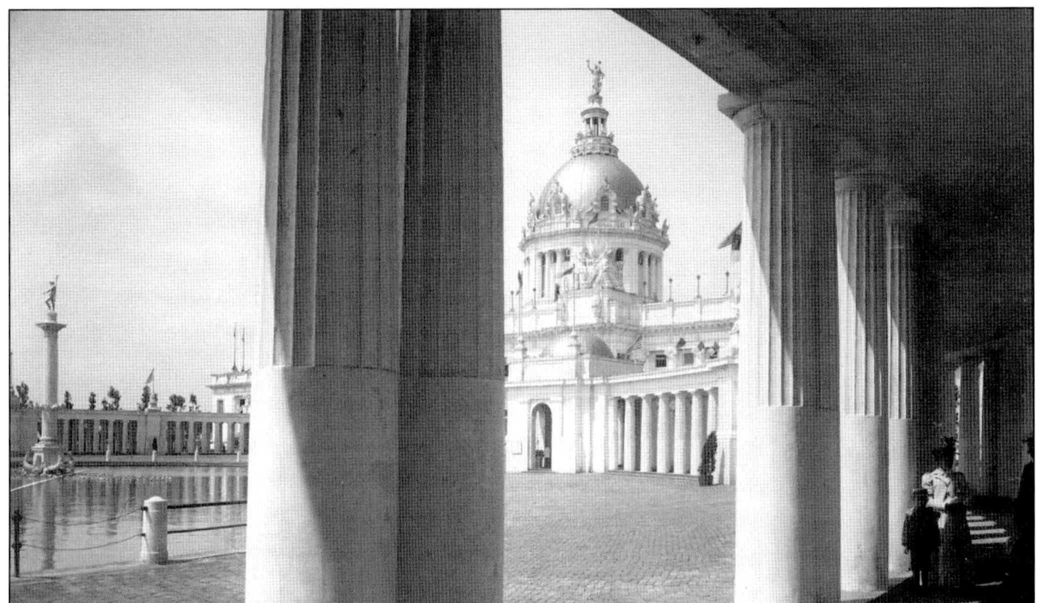

The grand scale of the United States Government Building is apparent in this view from the north colonnade. The three sections of the United States Government Building stretched a total distance of 504 feet. (Photo courtesy of Omaha Public Library.)

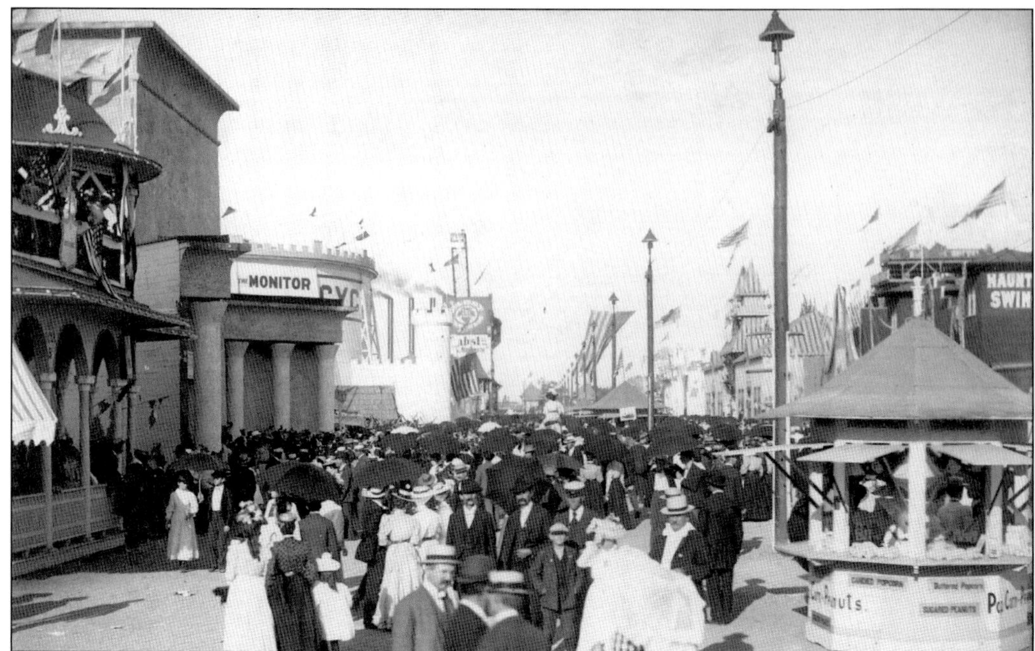

The intersection of the West and North Midways was a crossroads of activity at the exposition. The booth at the right sold candied or buttered popcorn and sugared peanuts. Note the young man pictured at the center, facing the camera. Clearly, this young exposition visitor recognized the historical significance of the event and wanted his attendance to be remembered by future generations. (Photo courtesy of Omaha Public Library.)

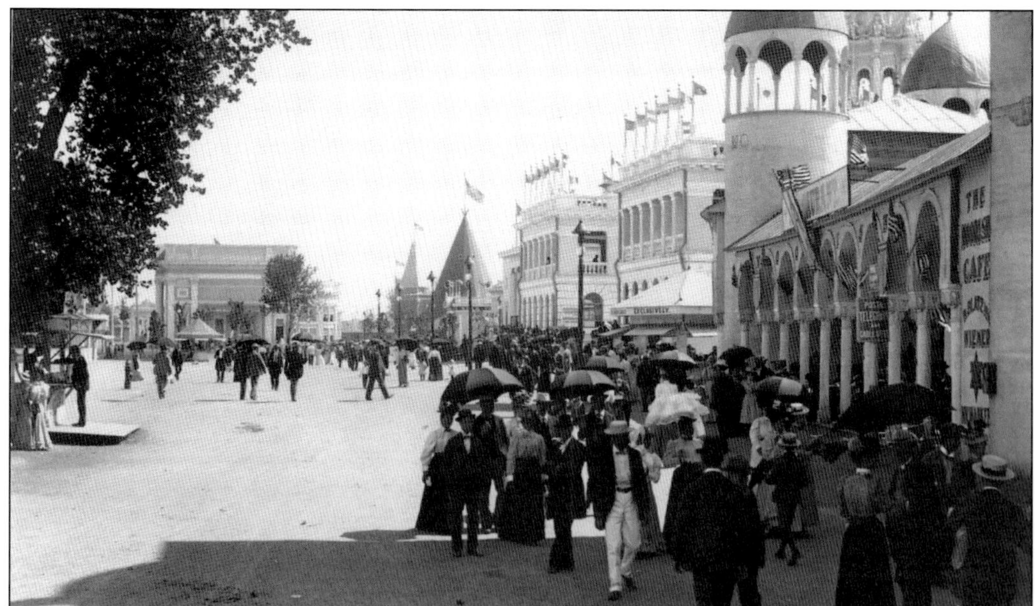

The East Midway of the exposition met the South Viaduct entrance of the Grand Court at the Grand Plaza. This area was the scene of considerable visitor interaction. Pictured at the right is the Moorish Palace and Café, an attraction that included a wax museum featuring likenesses of world leaders and other celebrities. (Photo courtesy of Omaha Public Library.)

Two

A Grand Achievement

Buildings dedicated to the pursuits of Agriculture, Fine Arts, Liberal Arts, Machinery and Electricity, Manufactures, Mining, and a seventh displaying the institutions of government, were at the heart of the Trans-Mississippi Exposition. Surrounding a central lagoon and paved walking path, the structures of the Grand Court housed displays from throughout the United States.

The buildings of the Grand Court were constructed of wood and covered in staff—a building material composed of plaster, cement, and a readily available fibrous material, such as hemp. Exposition organizers determined that all buildings would be of Classical design and possess a similar size, shape, and color. The critical success of the great "White City" at the Trans-Mississippi Exposition was the result of lessons learned from the efforts of previous expositions. During a building dedication speech, Trans-Mississippi Exposition President Gurdon W. Wattles remarked, "Who can stand at either end of the Grand Court and look at the magnificent spectacle of architectural grandeur there displayed without receiving impressions and inspiration which will last through life?"

Connecting all structures of the Grand Court were colonnades to provide covered passage between buildings and complete the stage upon which the exposition would be set. At the east end of the Grand Court were twin towers featuring two of the most popular restaurants at the exposition. Passing between the towers was the South Viaduct, which crossed Sherman Avenue, modern-day Sixteenth Street.

Memories have passed into history. The inevitable passage of time has changed the landscape. Only photographs of this grand achievement remain.

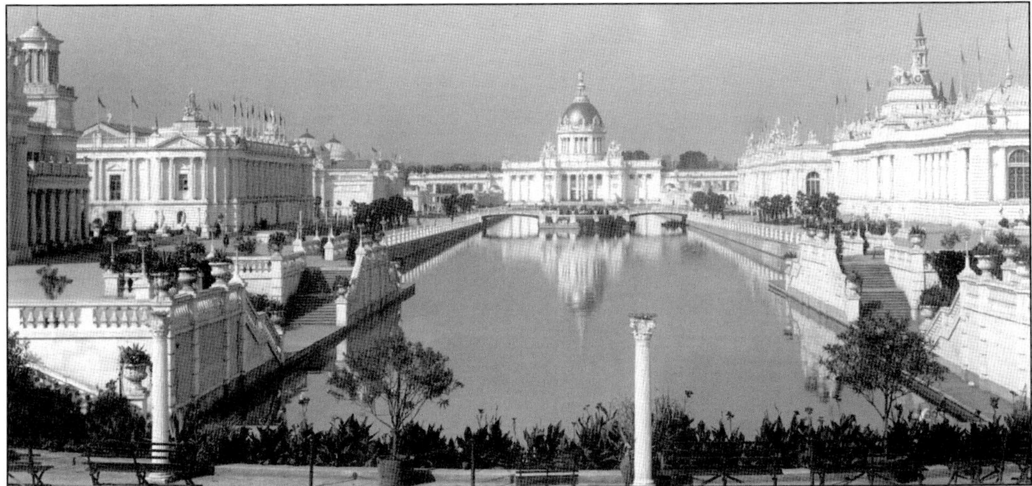

Golden sunshine and calm winds produced this view of the Grand Court, featuring a reflection of the United States Government Building in the exposition lagoon. The main buildings in this area of the exposition grounds were designed to be of a similar shape and height to enhance the grandeur of the setting. (Photo courtesy of Omaha Public Library.)

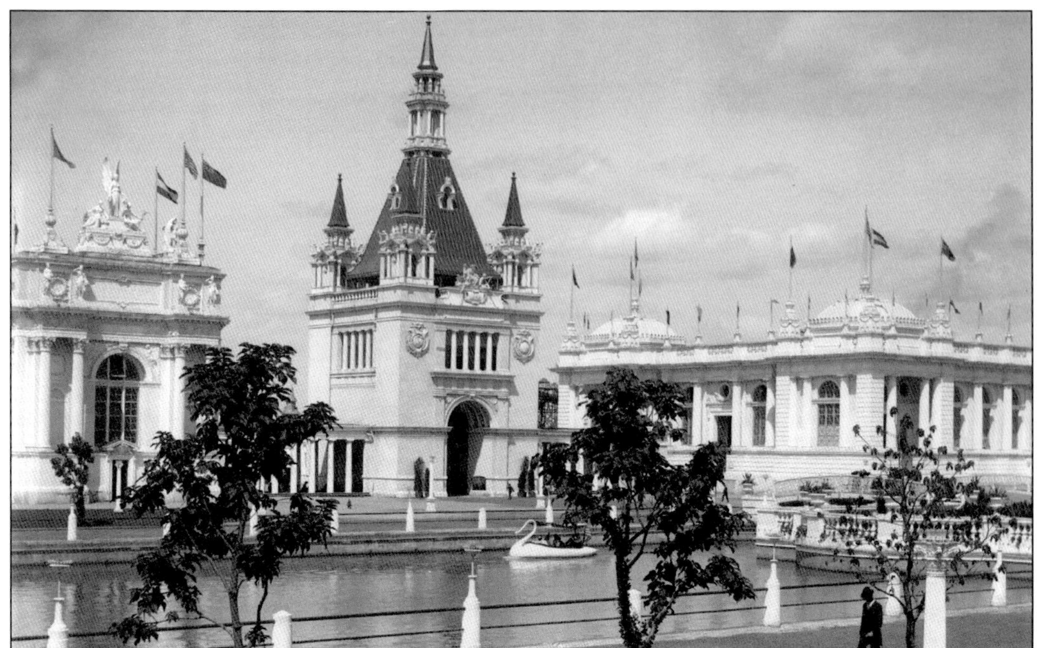

Though not the main entrance of the Trans-Mississippi Exposition, the Administration Arch was a focal point for visitors to the Grand Court. Facing the north side of the lagoon, this structure measured 50 by 50 feet at its base and reached a maximum height of 150 feet. (Photo courtesy of Omaha Public Library.)

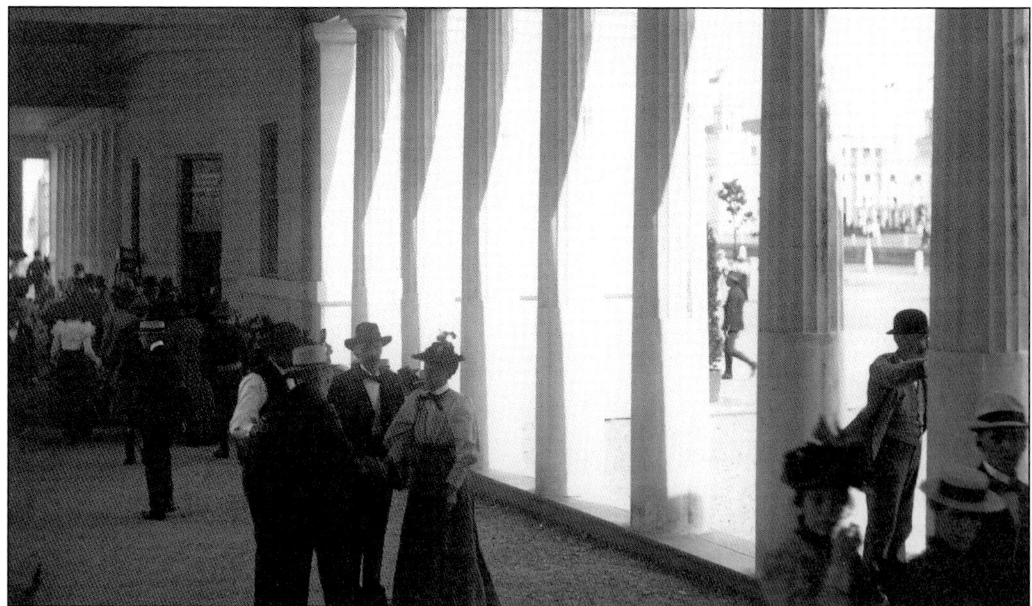

Exposition visitors meet inside the colonnade beneath the Administration Building. The Administration Building, including its central arch, provided passage between the Grand Court and the West Midway. The doorway pictured beyond the columns was for the elevator used to reach the upper floors of the building and an observation level. (Photo courtesy of Omaha Public Library.)

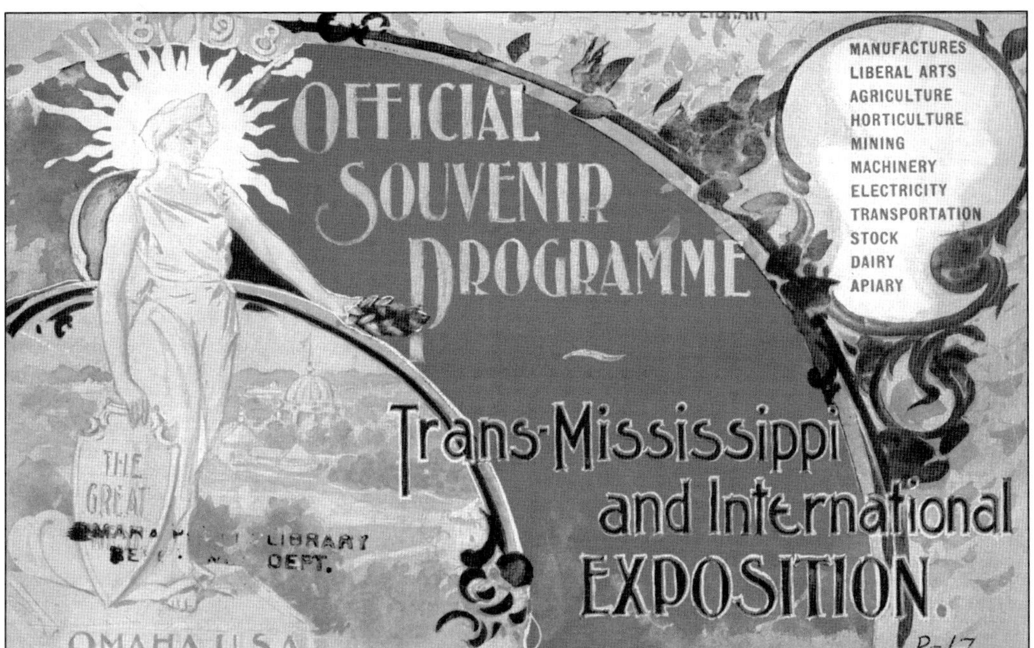

The official souvenir program of the Trans-Mississippi Exposition featured photographs of the exposition grounds, programs for daily events, and a list of conventions visiting Omaha for the week during which the program was purchased. (Photo courtesy of Omaha Public Library.)

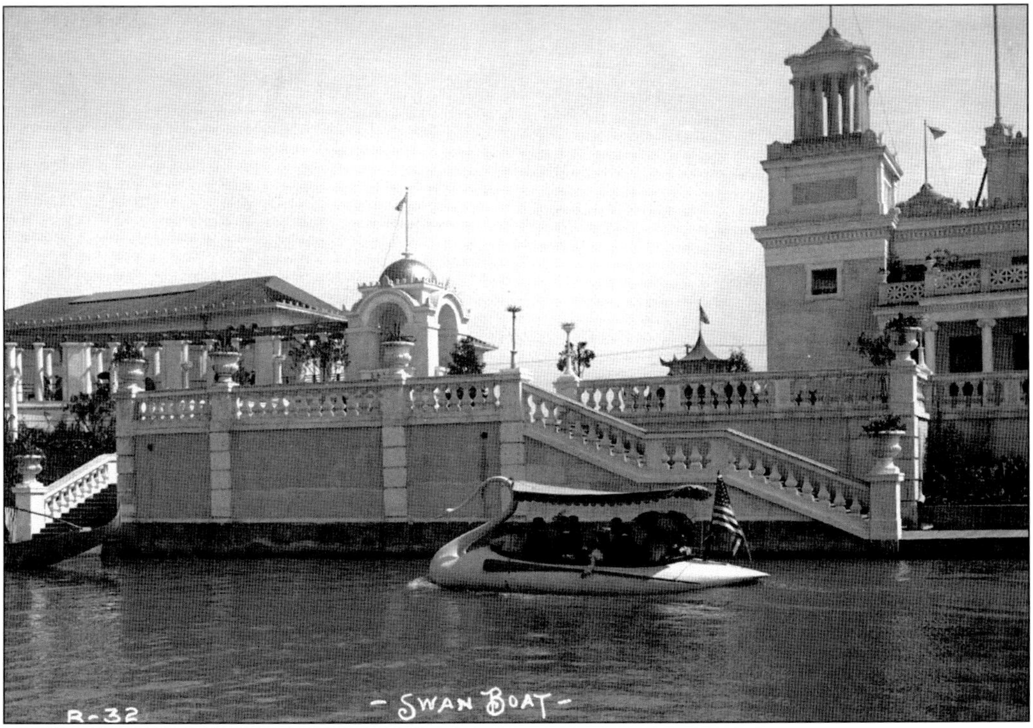

Swan boats treated visitors to a whimsical ride on the lagoon to better admire the architecture of the Grand Court. (Photo courtesy of Omaha Public Library.)

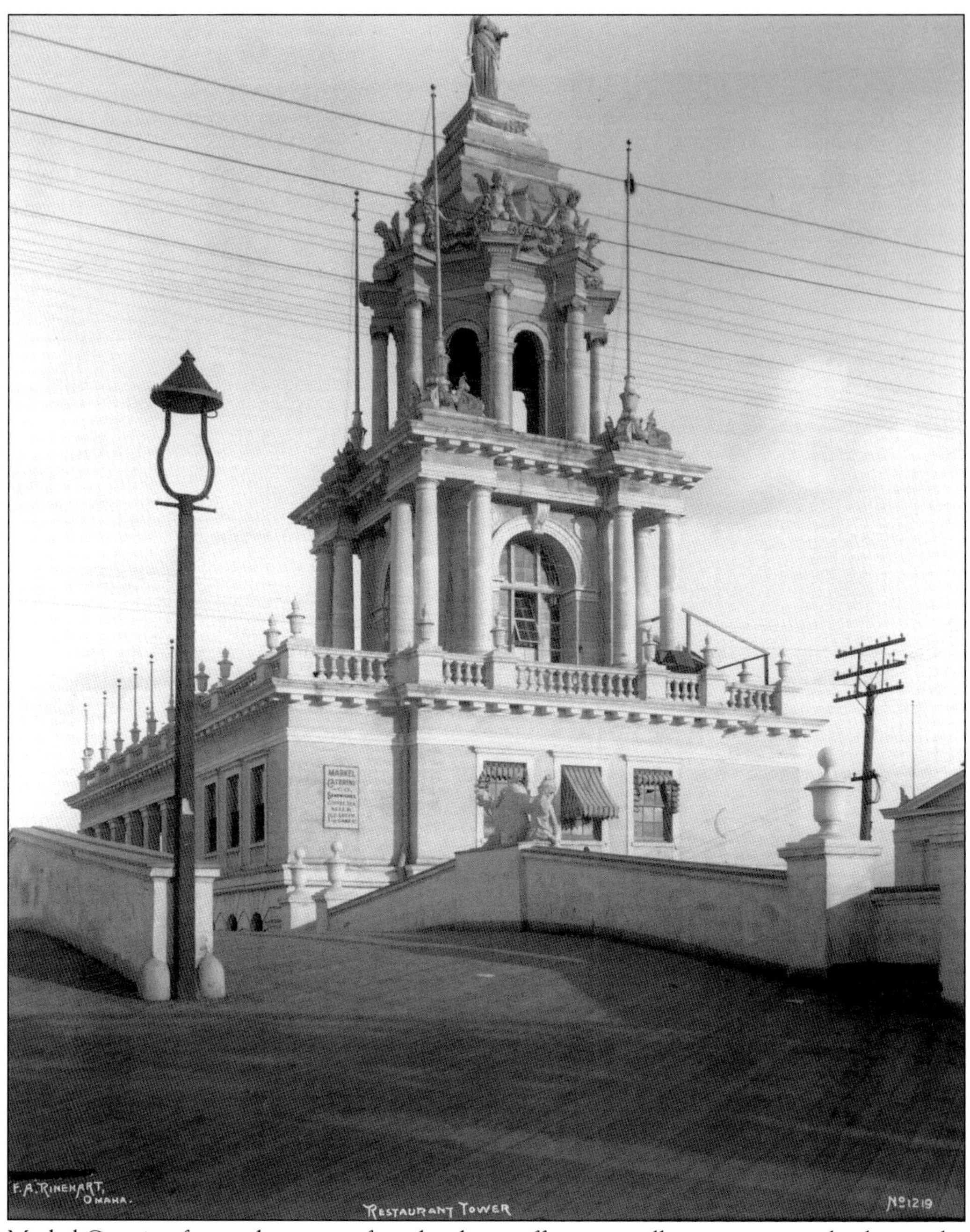

Markel Catering featured a menu of sandwiches, coffee, tea, milk, ice cream, and cakes in the restaurants of the towers at the South Viaduct of the exposition. Markel prepared meals for special occasions as well as meals for the many dignitaries who visited the exposition. (Photo courtesy of Omaha Public Library.)

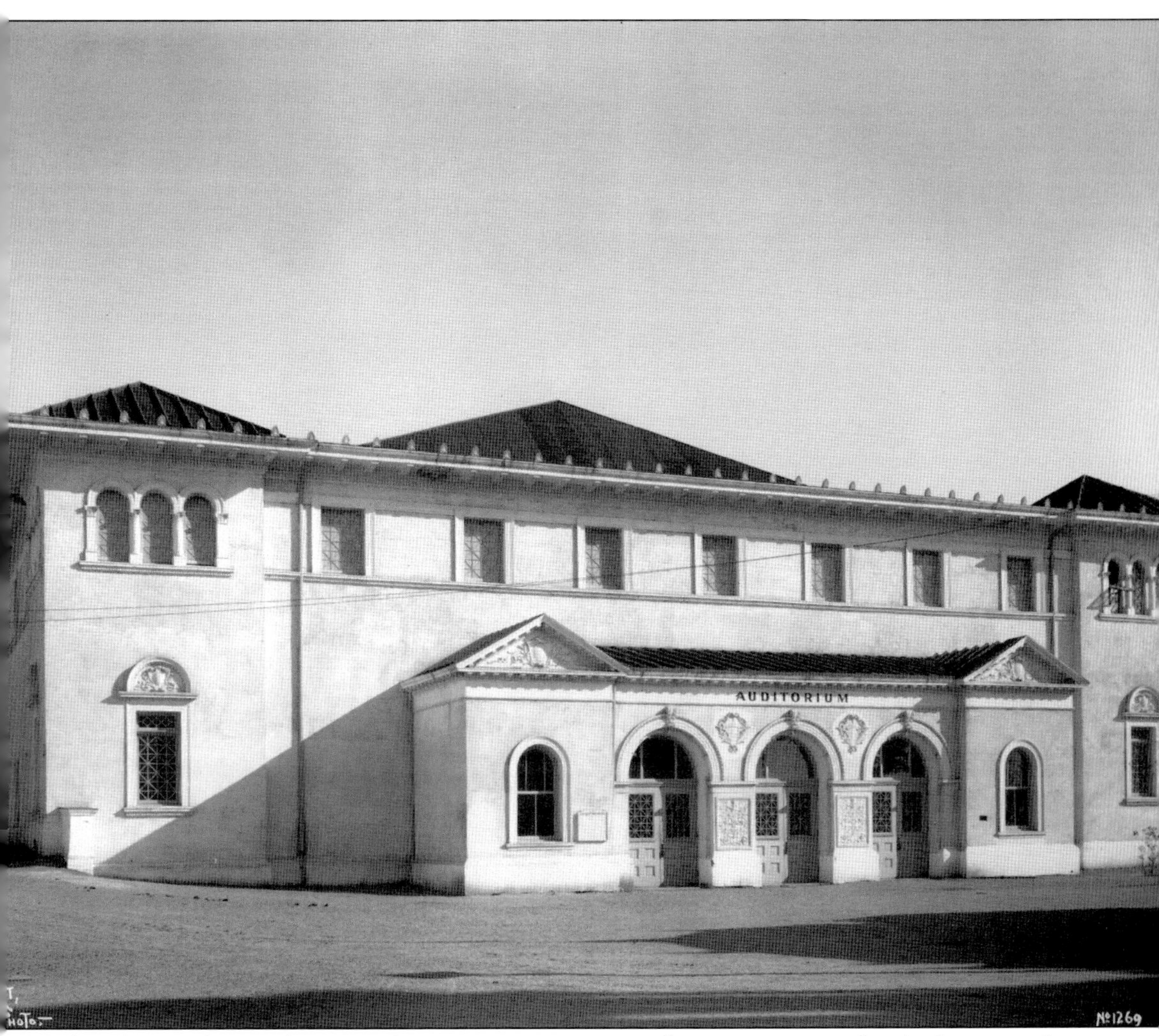

With a theater designed to provide seating in a semi-circular arrangement, the Auditorium of the Trans-Mississippi Exposition was the site of concerts, speeches, "Special Day" celebrations, delegate meetings, religious services, and organ recitals, which occurred daily at 1:30 p.m. (Photo courtesy of Omaha Public Library.)

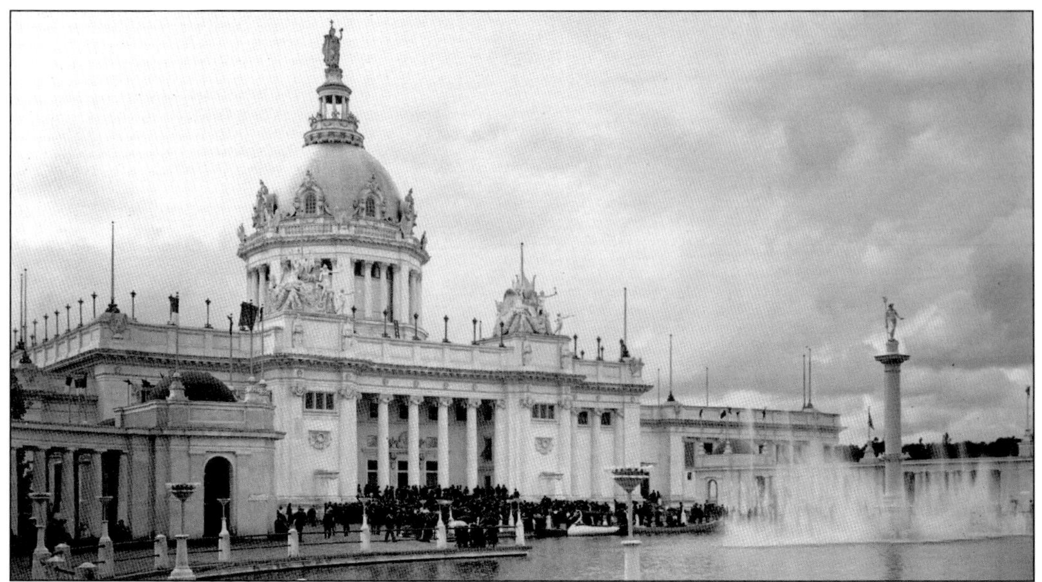

Facing the mirror basin was the 208-foot center section of the United States Government Building. At the top of the dome, 178 feet above the walking path below, was a reproduction of Frederic Bartholdi's "Liberty Enlightening the World," better known as the Statue of Liberty. Approximately 46,000 square feet of exhibition space included items such as the desk from which Thomas Jefferson penned the Declaration of Independence, signatures of all but two Presidents of the United States, and examples of every coin minted by the United States government. (Photo courtesy of Omaha Public Library.)

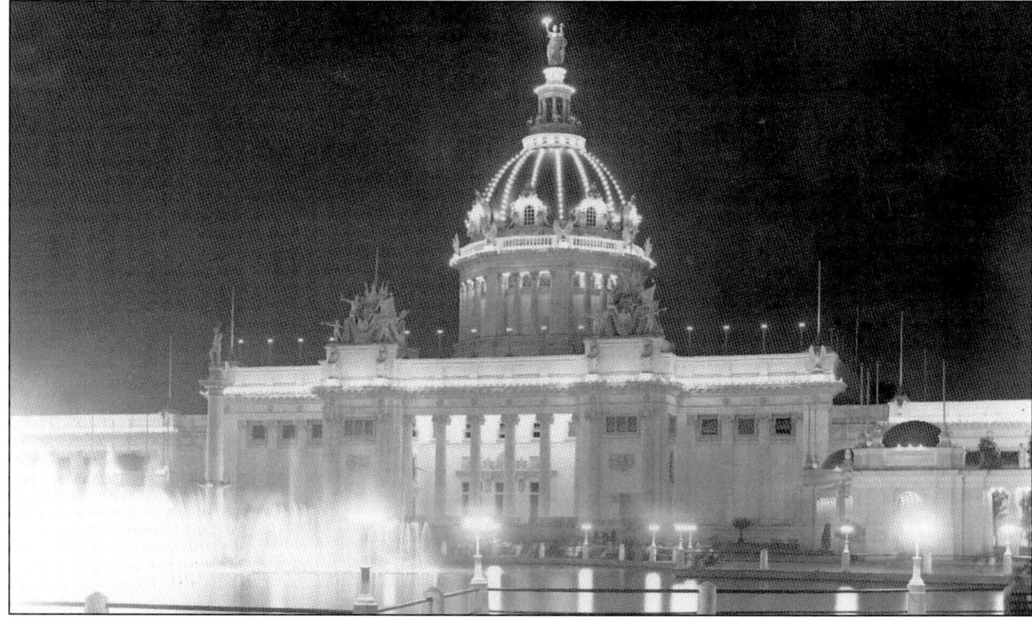

From across the mirror basin at the west end of the Grand Court, the illuminated spray of the fountain casts a glow on the United States Government Building. This unique lighting display provided a spectacular setting for nighttime events in front of the government building. (Photo courtesy of Omaha Public Library.)

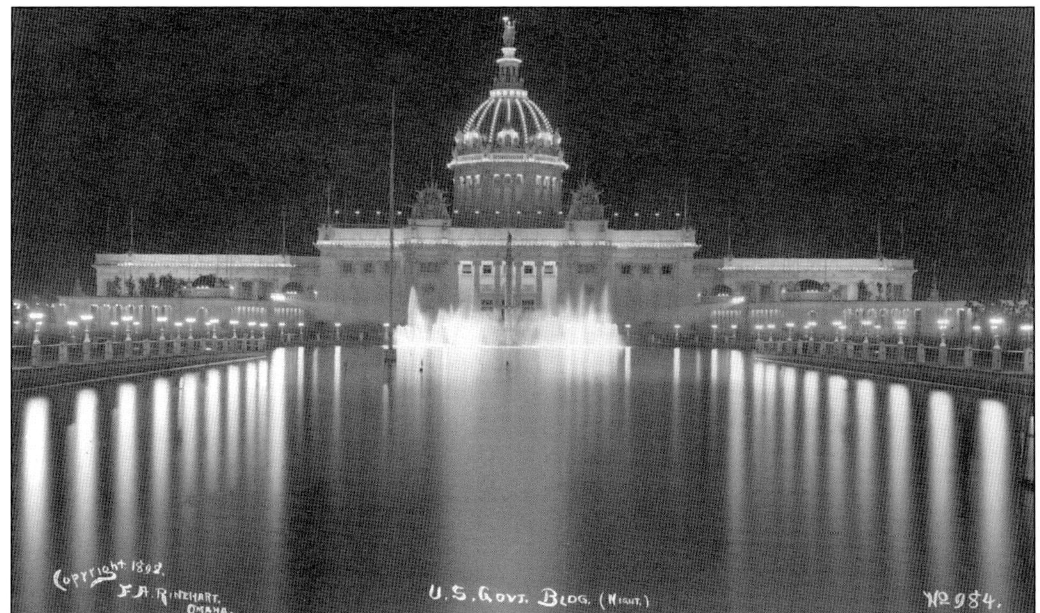

The lagoon and mirror basin created a spectacular effect when viewed against the backdrop of the illuminated buildings of the Trans-Mississippi Exposition. The United States Government Building was a focus of visitor attention for those who remained on the exposition grounds for this evening display. (Photo courtesy of Omaha Public Library.)

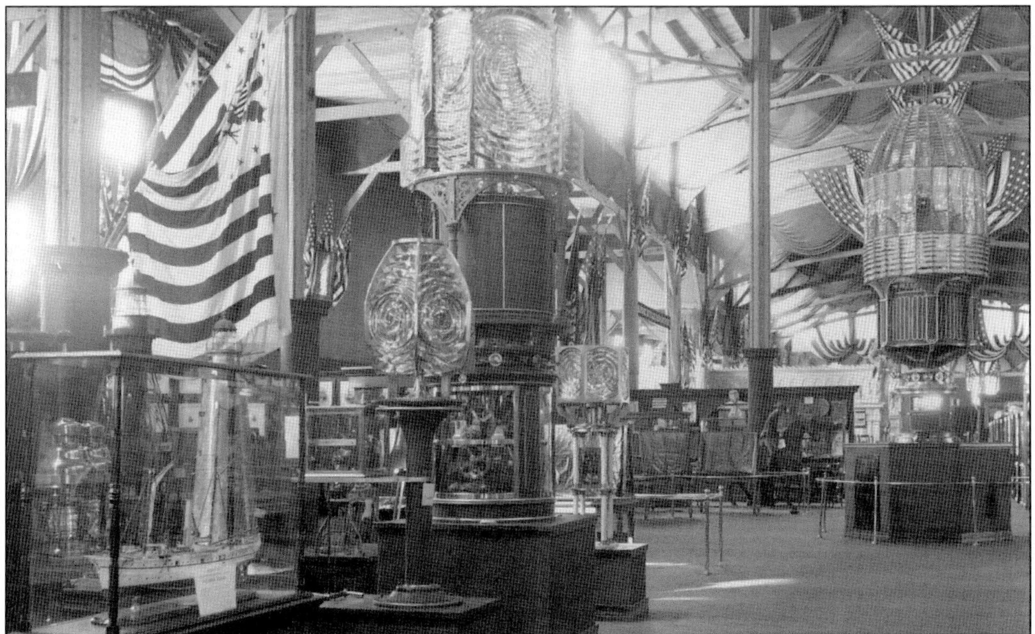

Lighthouse lenses, lanterns, and other associated items are pictured in this display from the United States Government Building. The display of the United States Postal Service is in the distance. A complete series of United States postage stamps, a sizable collection of foreign stamps, and a working post office were included in the postal service exhibit, as well as a display featuring the commemorative Trans-Mississippi postal issue. (Photo courtesy of Omaha Public Library.)

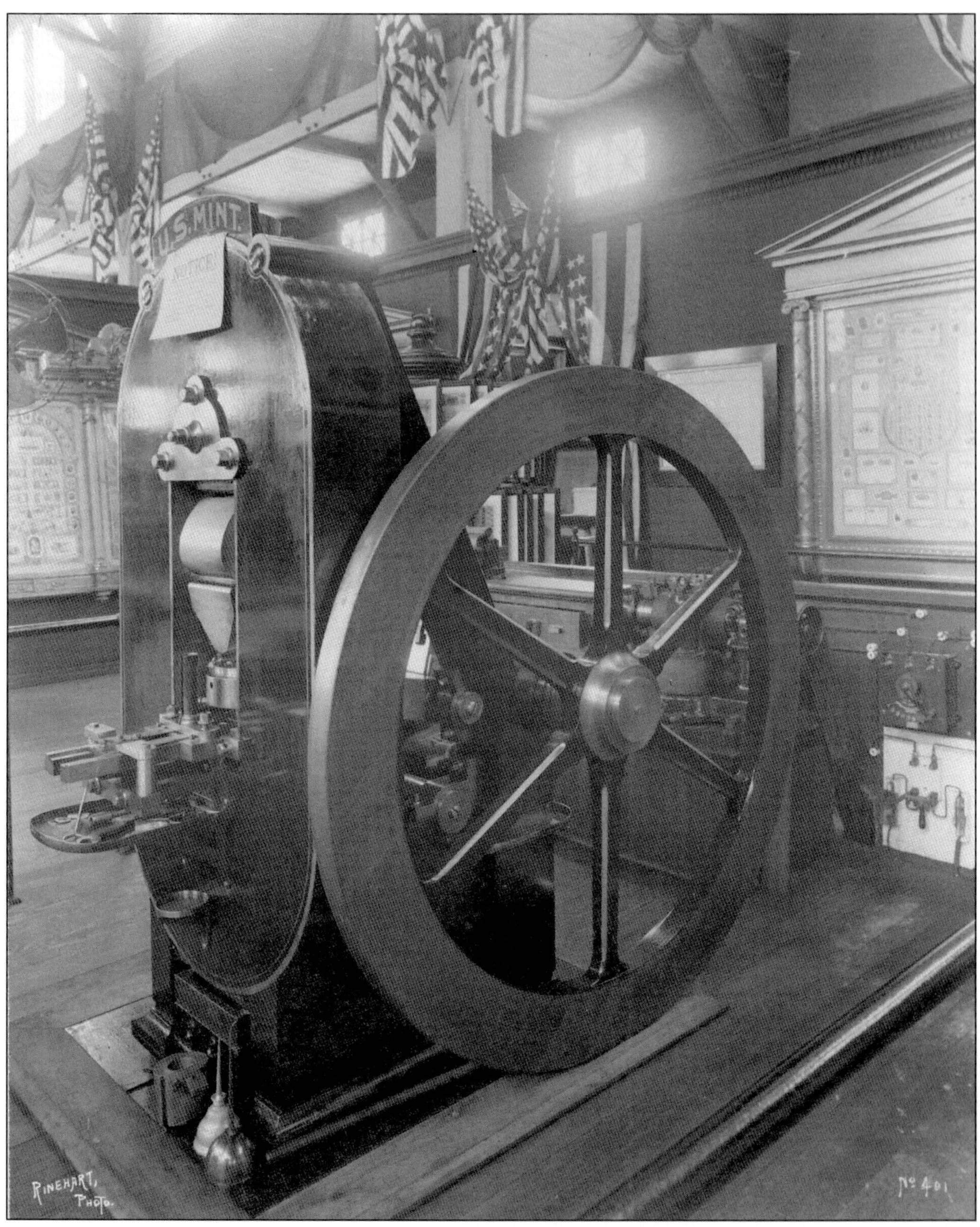

A coin press from San Francisco was used to create souvenir exposition medals. Approximately 25,000 souvenir medals were coined by this press during the exposition season. This display in the government building fascinated many visitors, who watched as the medals were struck and then purchased this definitive exposition souvenir at outlets throughout the exposition grounds. (Photo courtesy of Omaha Public Library.)

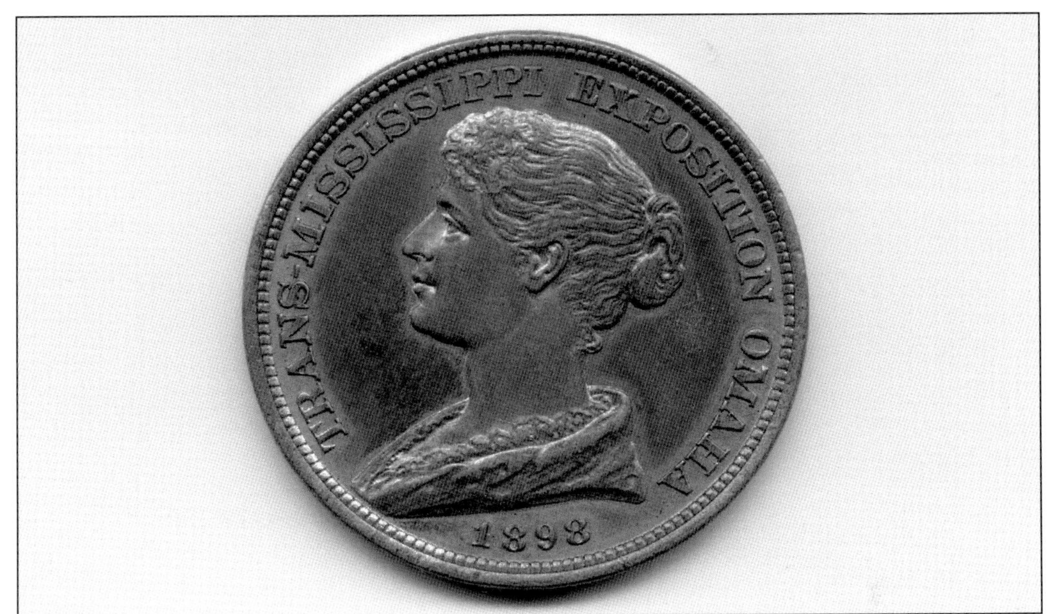

The bust appearing on the obverse side (*above*) of the Trans-Mississippi Exposition souvenir medal was developed from a composite of photographs taken of two women from each of the 22 Trans-Mississippi states. Curiously, the final composite photograph used to design the medal was developed by an artist from New York City. The reverse side (*below*) of the medal features a Native American in the act of spearing a buffalo and the date "1848," celebrating fifty years of the American West.

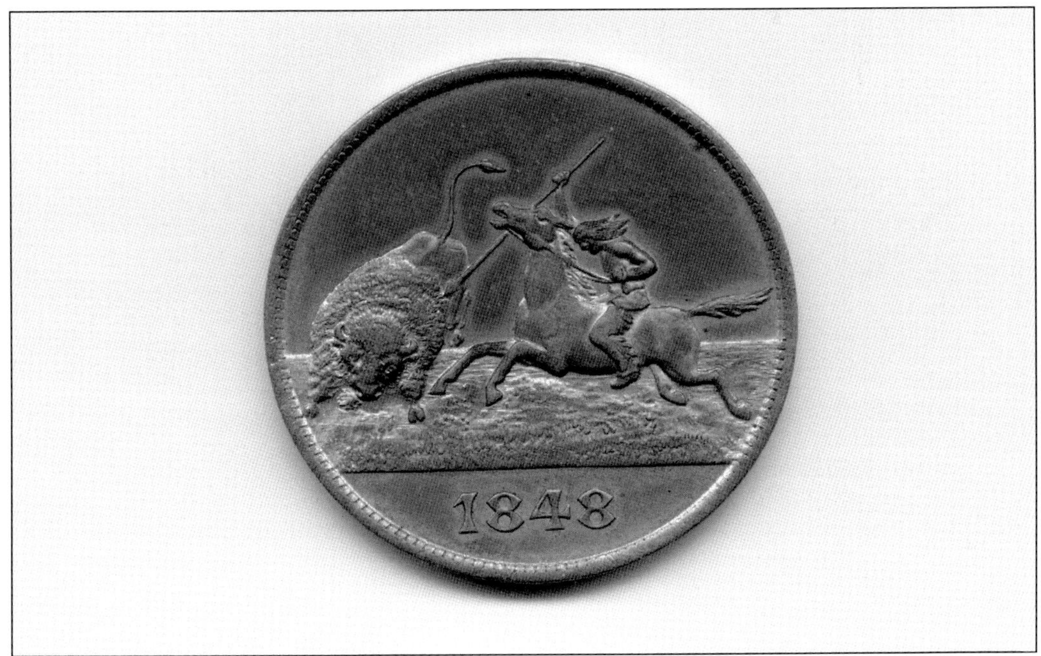

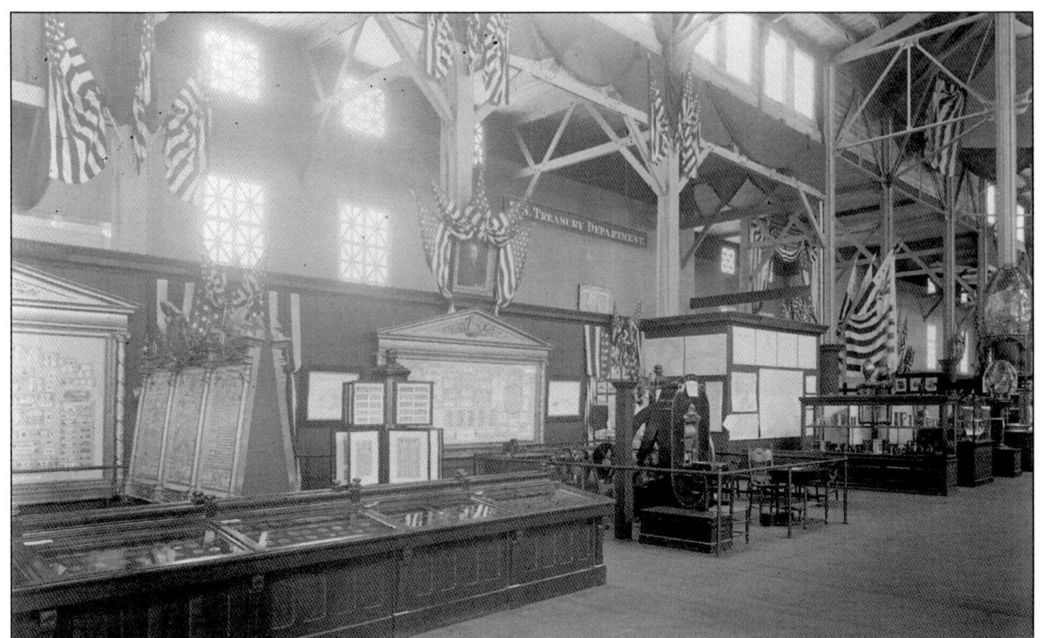

The display of the United States Treasury Department included a variety of paper currency, as well as postage and revenue stamps. Also on display were the wildcat bank notes and Confederate notes of a bygone era. Approximately 4,200 square feet of floor space was used for the Treasury Department exhibit. (Photo courtesy of Omaha Public Library.)

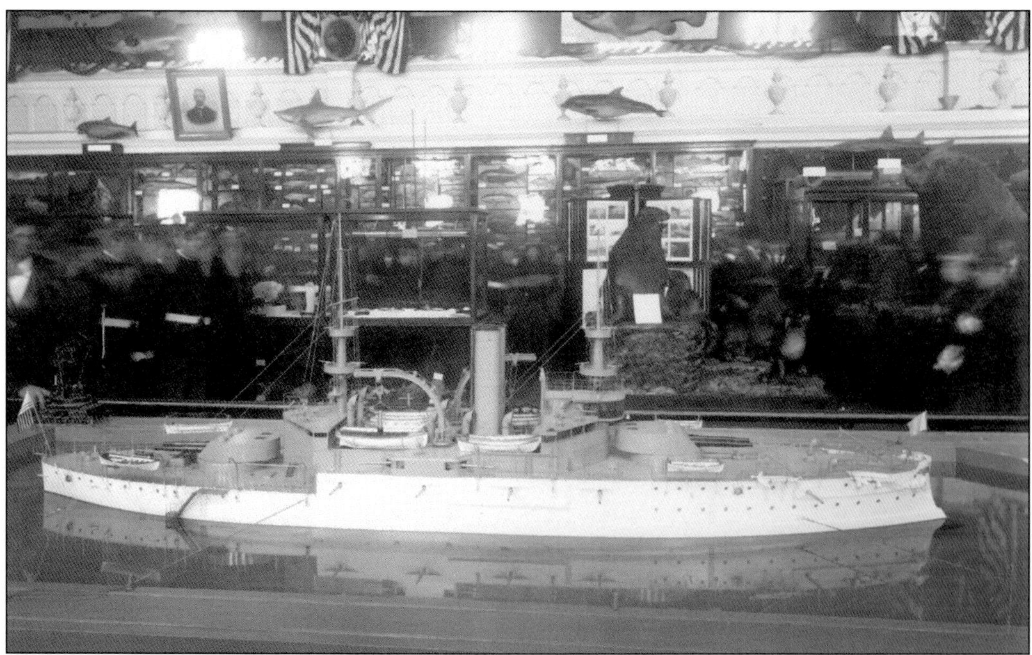

The battleship *Illinois*, a scale model of which was featured in an exhibit by the Department of the Navy, was a new addition to the fleet of Navy fighting vessels. A demonstration of battleship docking procedures and a detailed description of the battleship were provided at scheduled intervals to interested visitors. (Photo courtesy of Omaha Public Library.)

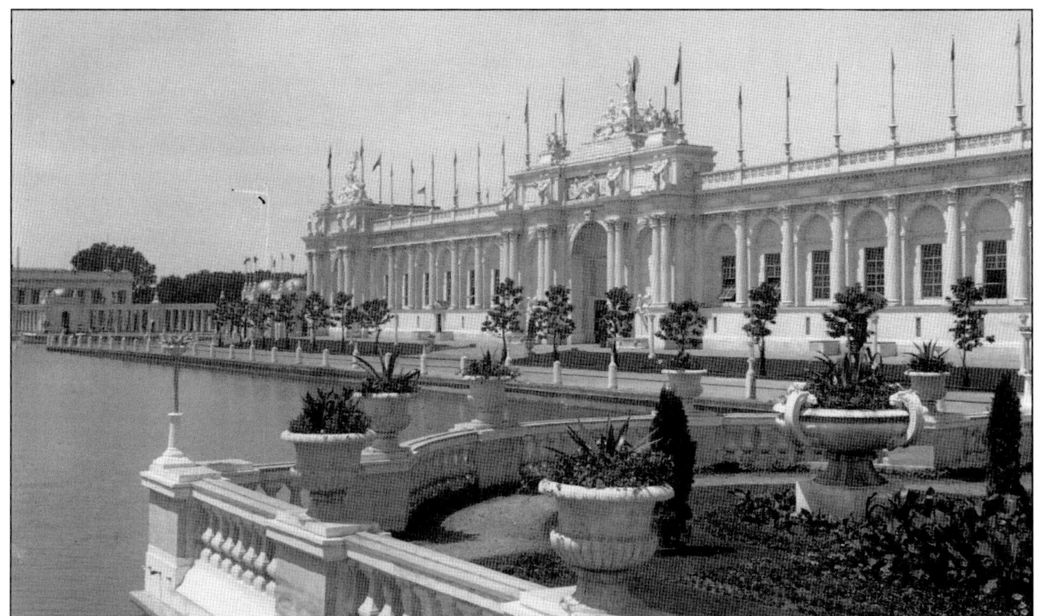

The trees, shrubbery, vines, and flowers used to decorate the Grand Court were a key component that, according to exposition planners, were "the means of shutting out a triste and sordid background" when combined with the buildings and connecting colonnades. The Agriculture Building is pictured from the bridge crossing the lagoon. (Photo courtesy of Omaha Public Library.)

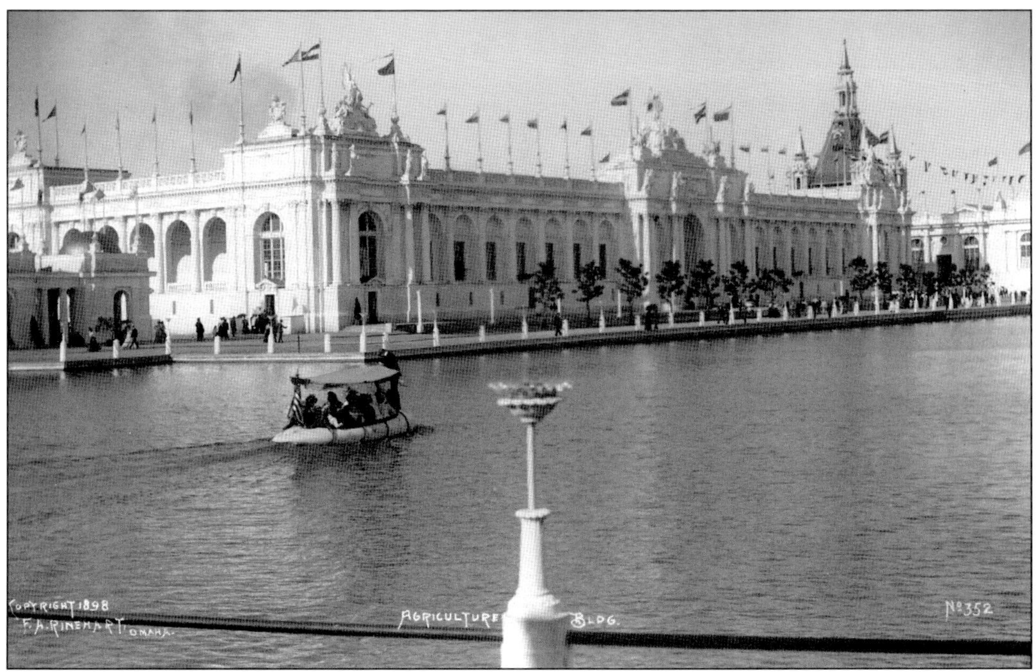

The Agriculture Building featured 84,260 square feet of floor space. Farm products from nearly all Western states were arranged for decorative and practical display. An exhibition promoting the achievements of agriculture was welcomed by a region of farmers who had recently experienced drought and plummeting farm prices. (Photo courtesy of Omaha Public Library.)

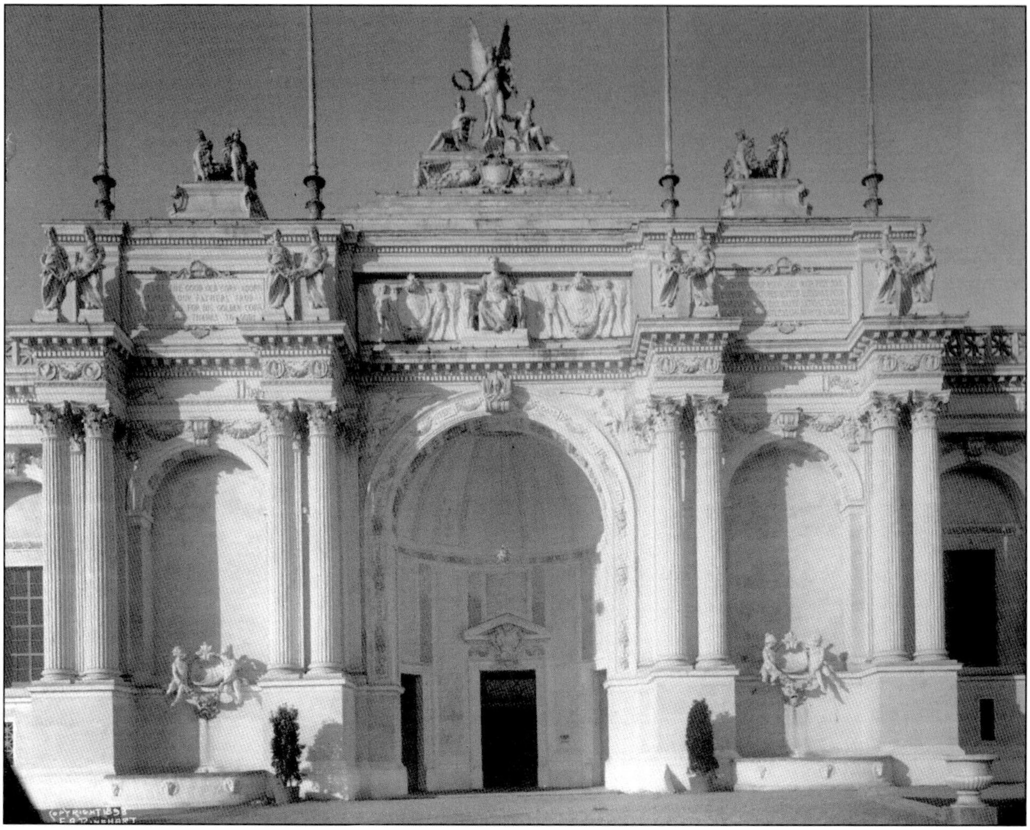

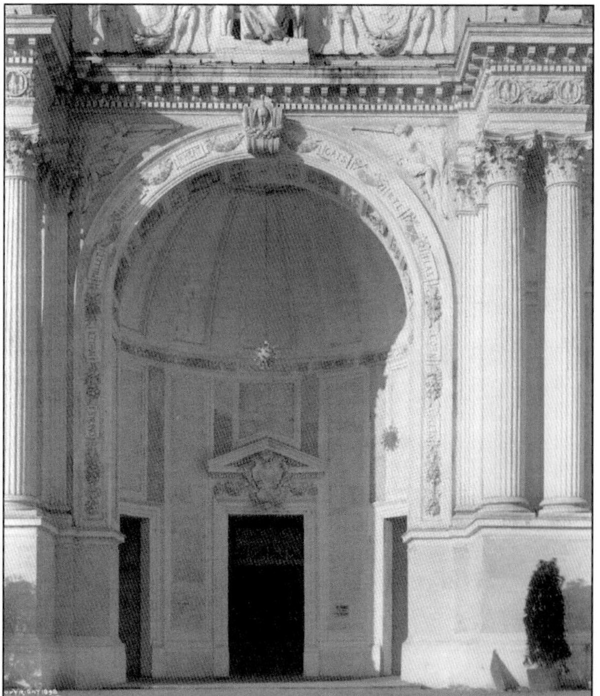

The main entrance of the Agriculture Building featured the following two inscriptions:
"But let the good old corn adorn the hills our fathers trod. Still let us for his golden corn send up our thanks to God!"
"With superior boon may your rich soil exuberant natures better blessings pour o'er every land, the naked nations clothe, and be the exhaustless granary of a world."

A lighted ball hangs from the top of the vestibule leading to the entrance of the Agriculture Building. Surrounding the arch, among the ornate decorations of this entrance, are the names of farm products on display inside: onions, barley, millet, buckwheat, wheat, oats, rye, flax, corn, and beets. (Photos courtesy of Omaha Public Library.)

Represented on the facade of the Agriculture Building was the bounty of farms spread throughout the Trans-Mississippi region. Originally envisioned to feature the colors of nature, the ornamentation was left in the ivory color of materials used to construct and decorate the buildings of the Grand Court. (Photo courtesy of Omaha Public Library.)

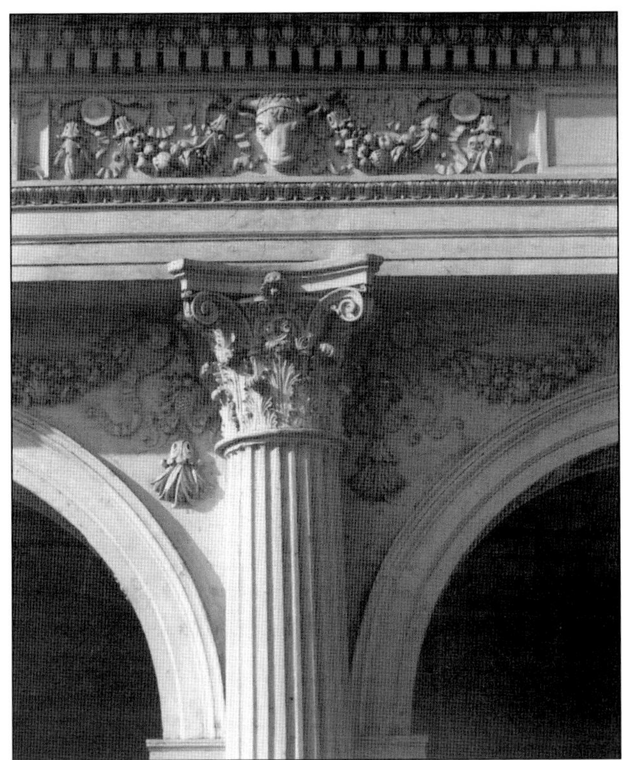

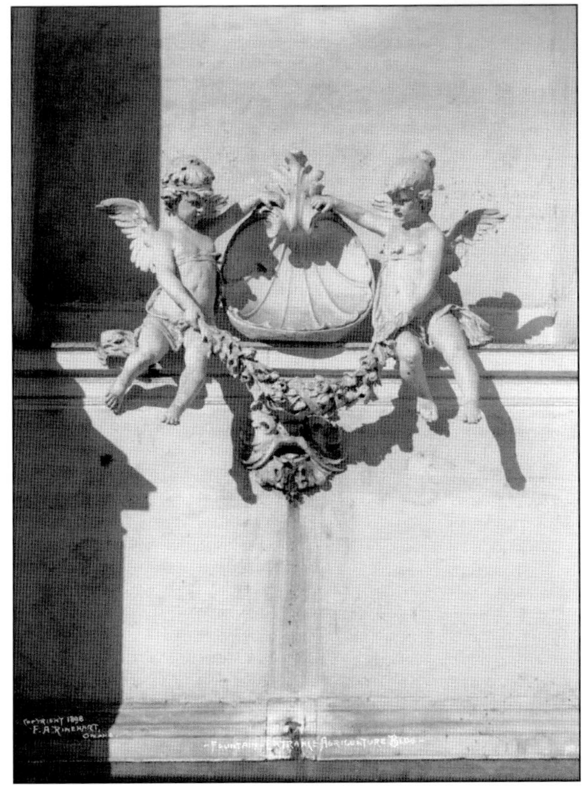

Two cupids support a basin above a small pool at the entrance of the Agriculture Building. Similar ornamental fountains were placed on either side of the entrance. (Photo courtesy of Omaha Public Library.)

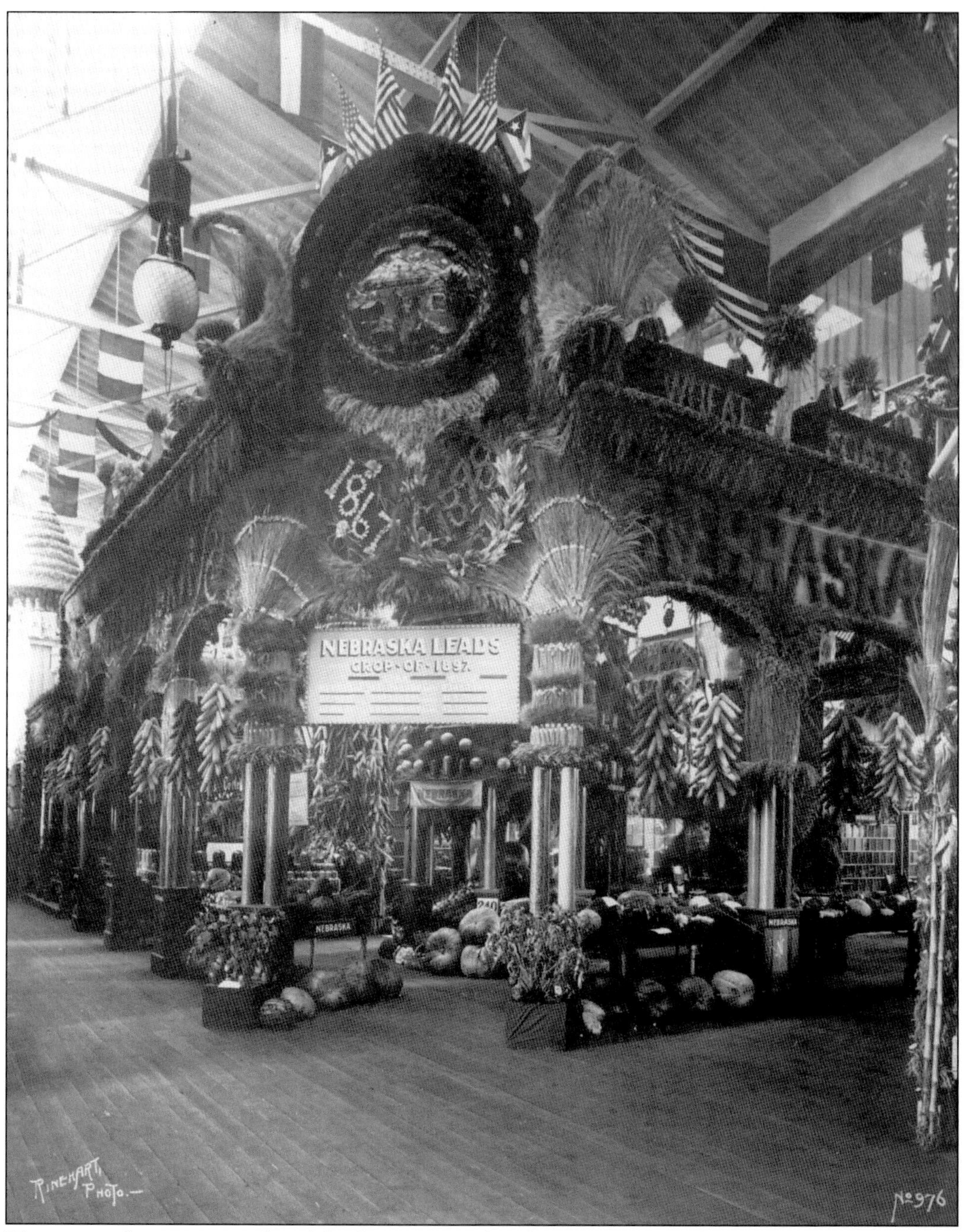

This comprehensive agricultural display attempted to draw attention to the farm products harvested in Nebraska in 1897. Beneath the flags at the top of this display is a large-scale depiction of the State of Nebraska seal created from seeds of various colors. The predominant color scheme of this entire exhibit included greens, reds, and yellows, the colors of a local philanthropic organization, the Knights of Ak-Sar-Ben. (Photo courtesy of Omaha Public Library.)

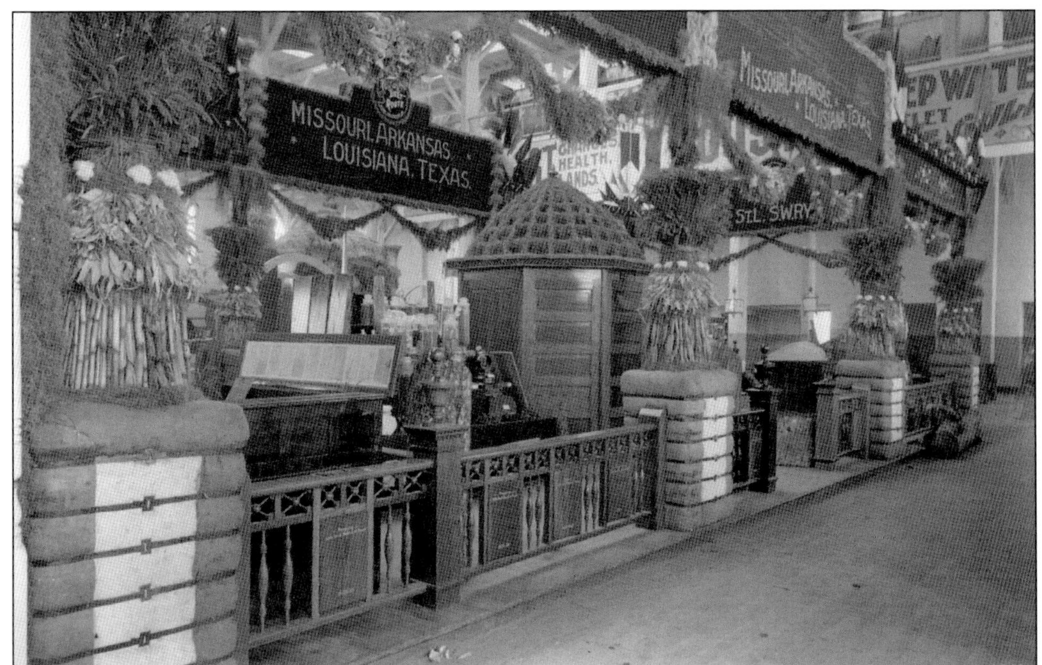

A gold medal was awarded to the St. Louis Southwestern Railway for its "Cotton Belt Route" display in the Agriculture Building. The display featured the agricultural and timber products of the areas serviced by the railway in Missouri, Arkansas, Louisiana, and Texas. The railing enclosing the exhibit was made from select varieties of wood found in this region. (Photo courtesy of Omaha Public Library.)

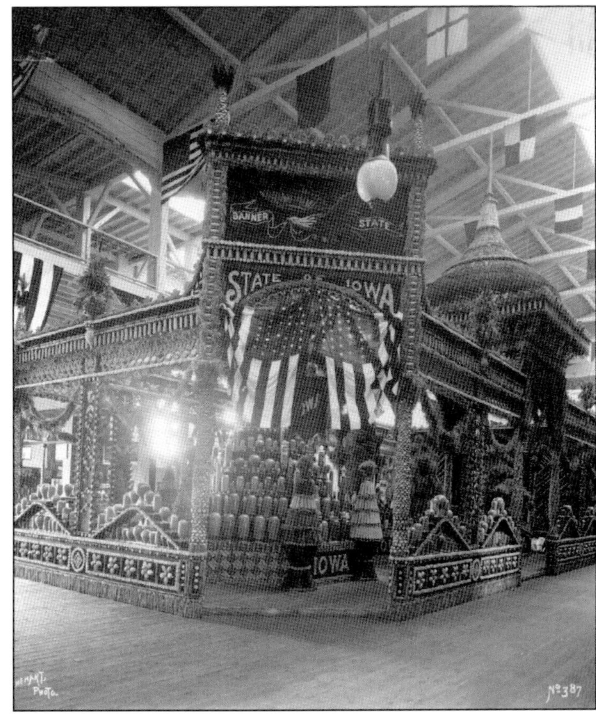

The State of Iowa presented one of the largest displays of agricultural products at the Trans-Mississippi Exposition. Featured in this exhibit were grains, seeds, corn, and a variety of other farm products from this Nebraska neighbor. (Photo courtesy of Omaha Public Library.)

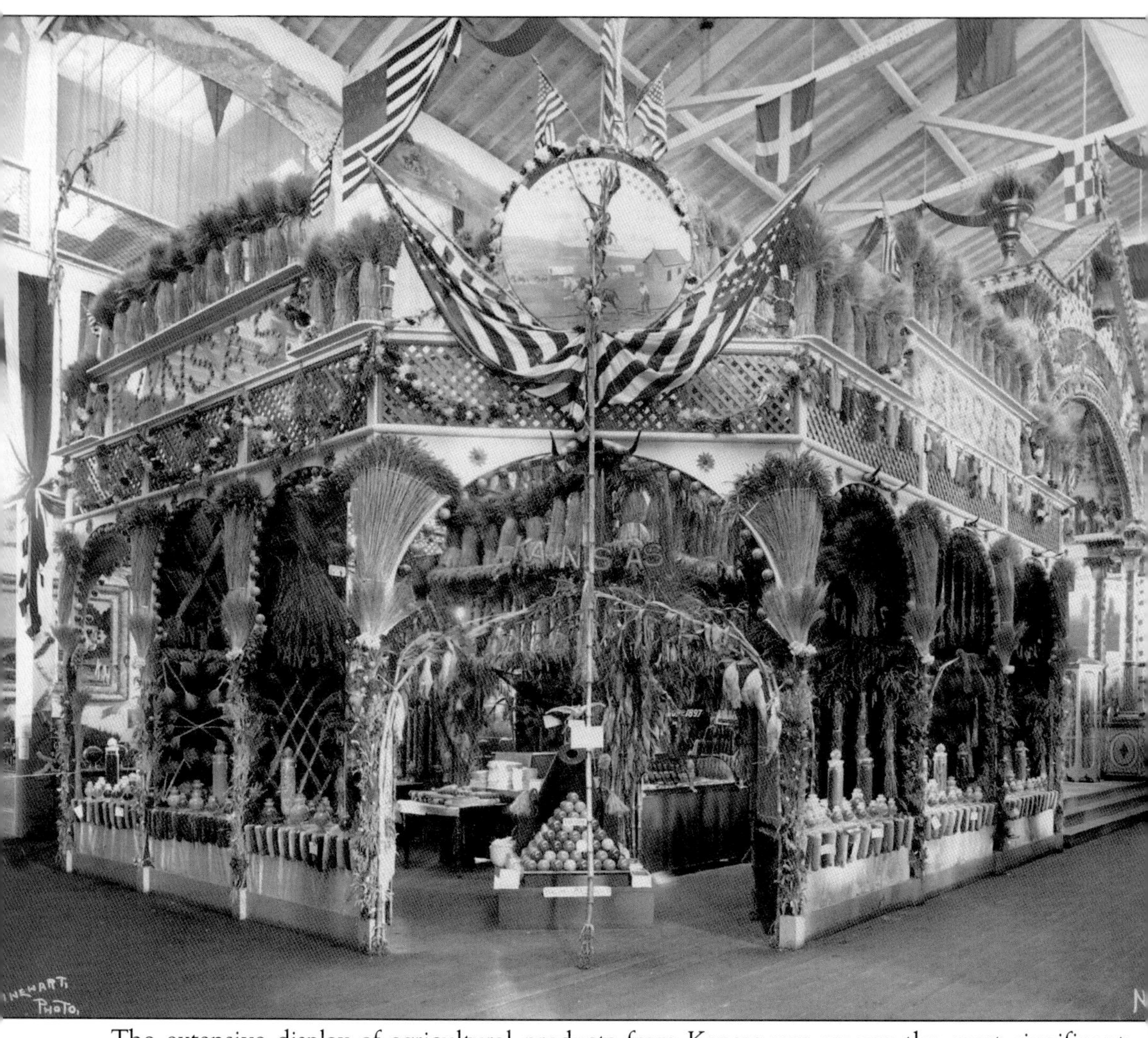

The extensive display of agricultural products from Kansas was among the most significant exhibits within the Agriculture Building. Featured products included apples, grains, and wood products from throughout the state. (Photo courtesy of Omaha Public Library.)

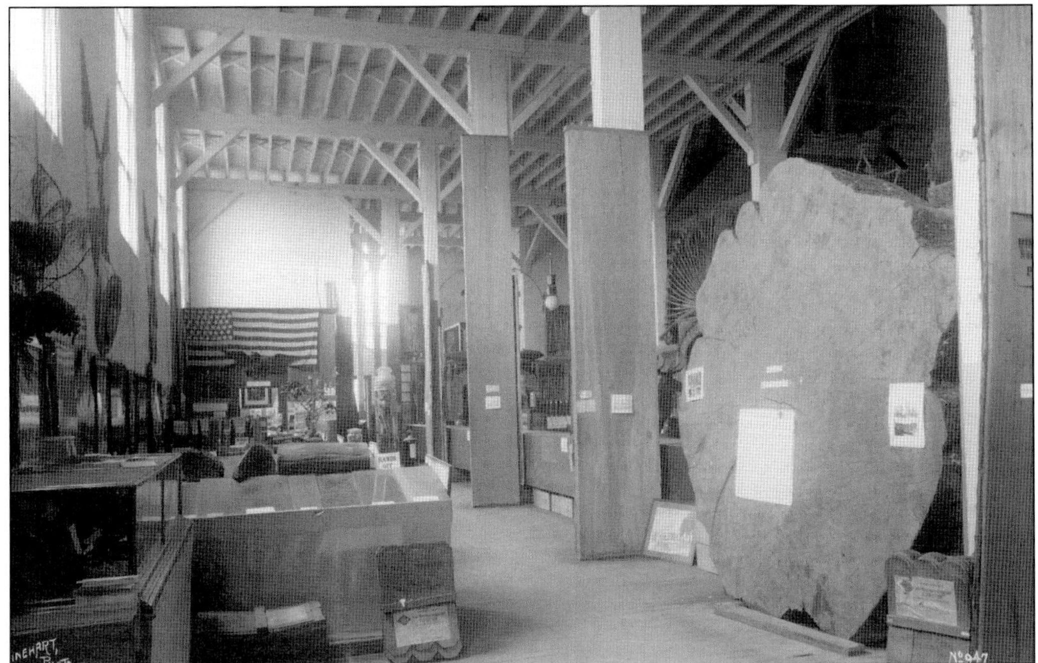

The forestry exhibit at the exposition included displays of White Pine, Bull Pine, White Fir, and Douglas Fir. Along with the large crosscut pictured here, scientific displays considered the quality of the wood along with accompanying forest foliage. A display from the state of Washington featured finished and raw lumber, including a single log that was 90 feet long and four feet in diameter. (Photo courtesy of Omaha Public Library.)

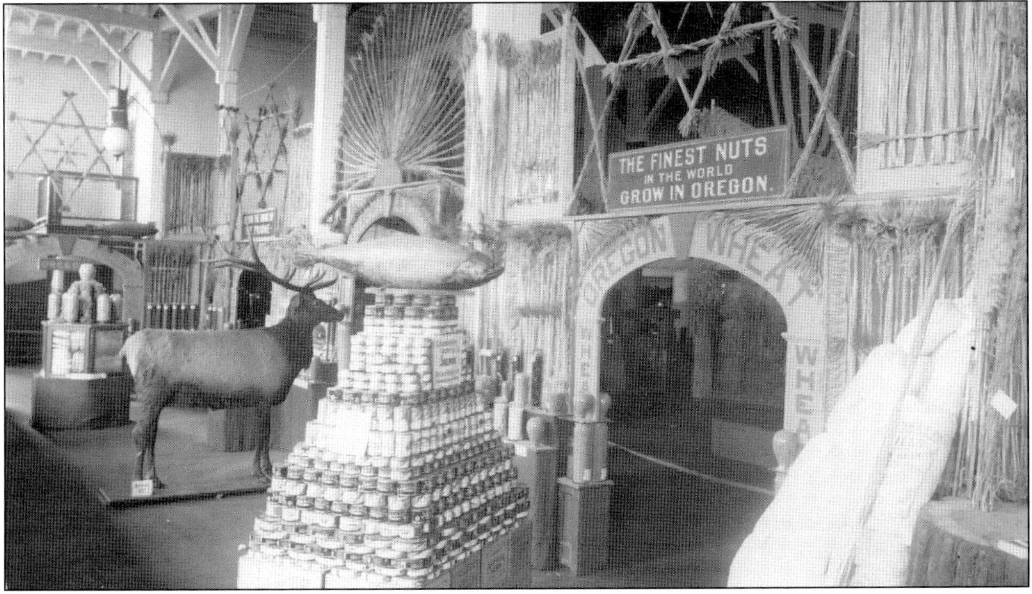

Along with "the finest nuts in the world," the agricultural display from the state of Oregon included examples of grains, fruit, timber, oats, and wheat. The unique display of canned salmon from the Columbia Canning Company is pictured in the foreground. (Photo courtesy of Omaha Public Library.)

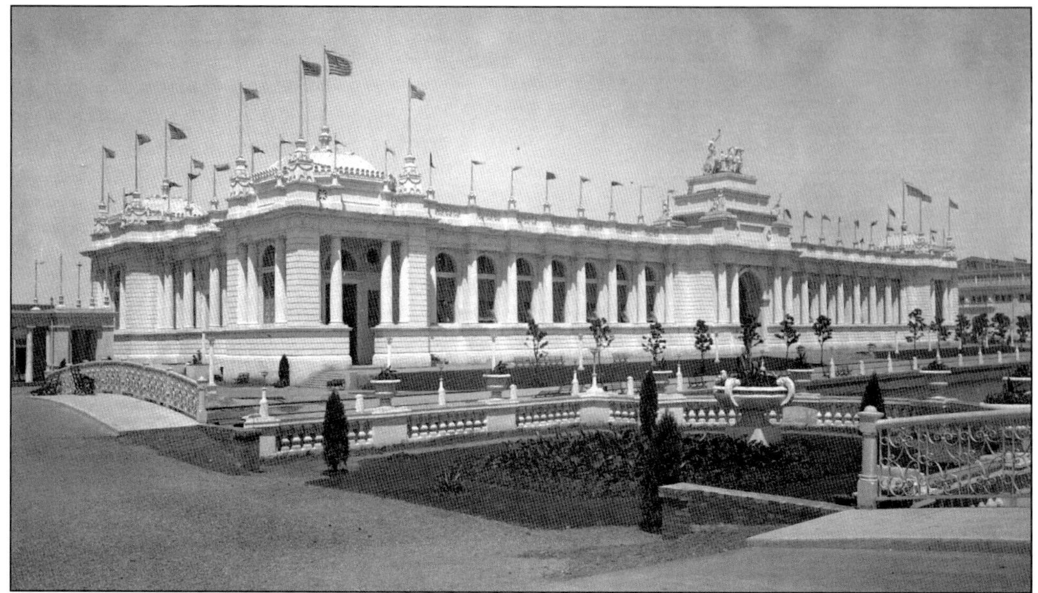

The 400-foot front of the Manufactures Building faced the lagoon from its north side. Sixteen feet separated the columns between the glass windows of the facade. The arch at the center of this structure spanned the main entrance at a height of 35 feet, welcoming visitors to a display of manufactured goods from throughout the world. (Photo courtesy of Omaha Public Library.)

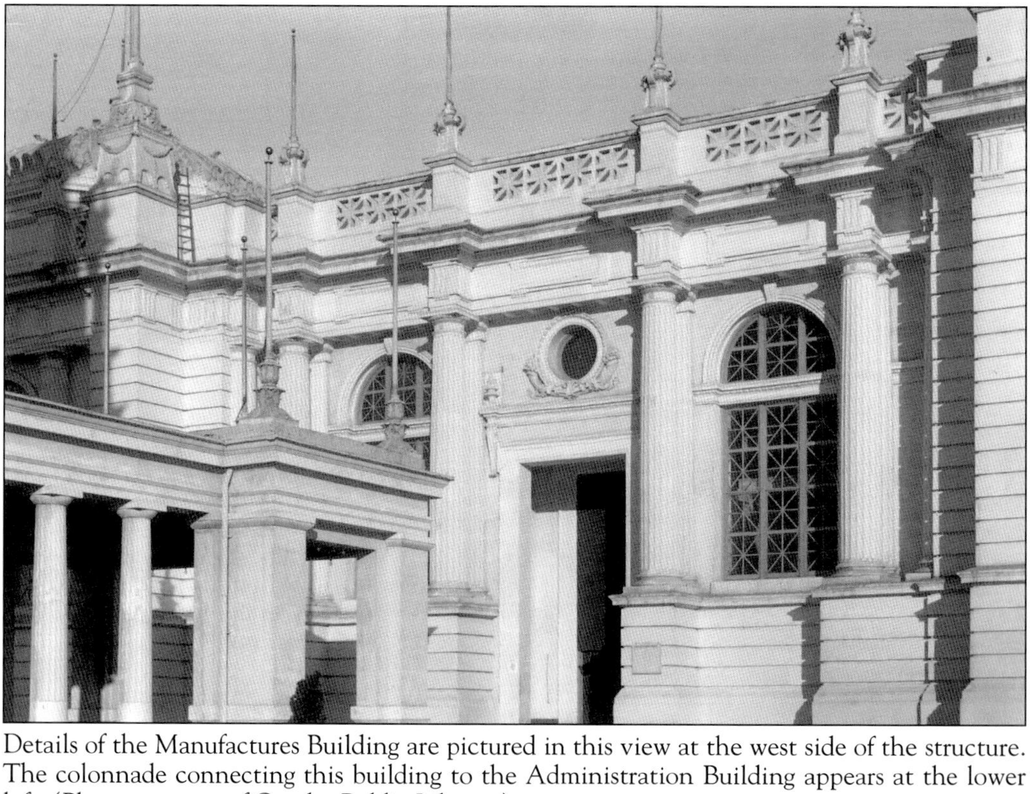

Details of the Manufactures Building are pictured in this view at the west side of the structure. The colonnade connecting this building to the Administration Building appears at the lower left. (Photo courtesy of Omaha Public Library.)

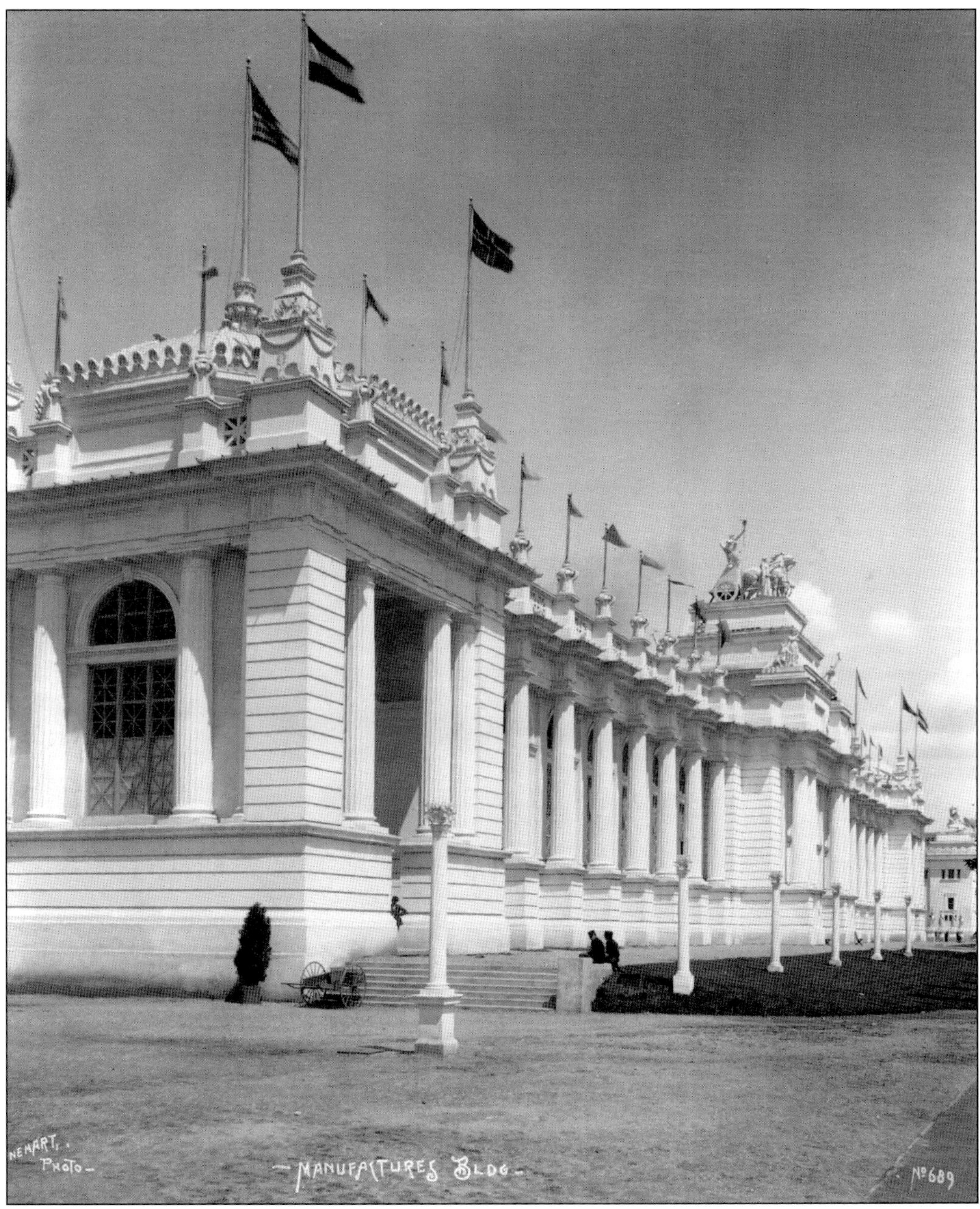

The southwest corner of the Manufactures Building shows the pathway leading to the West Midway. Included in this building among displays featuring a variety of practical and luxury goods was a chocolate manufacturer who provided free samples while orchestra musicians played from the second floor of the booth. (Photo courtesy of Omaha Public Library.)

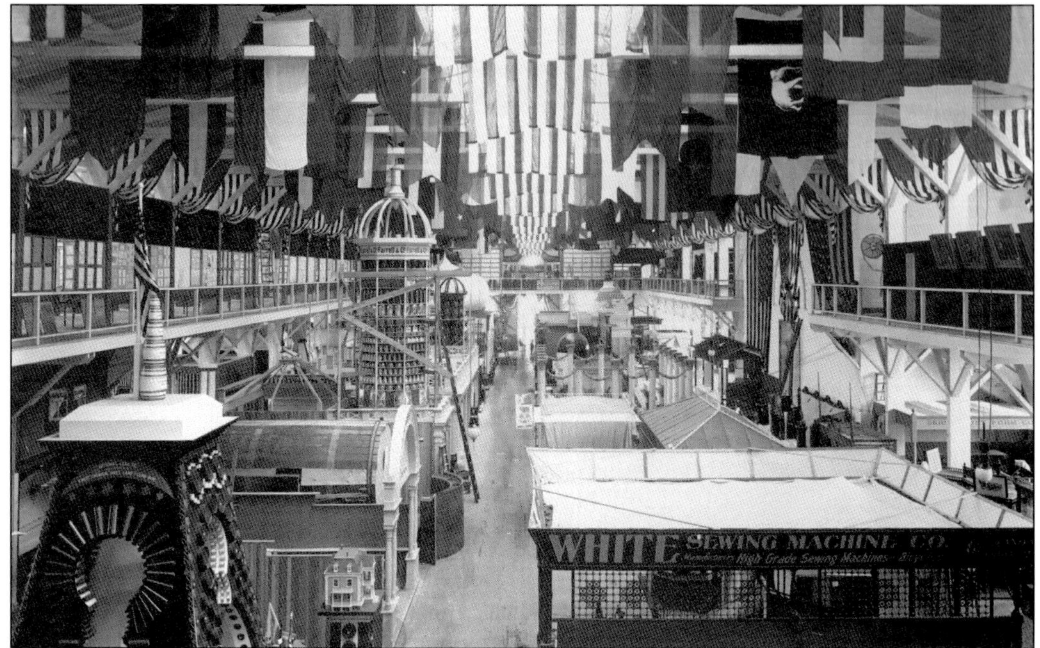

Among the exhibits of the Manufactures Building were many "living" exhibits. Demonstrations and motion were key components for this type of exhibit. Included in the Manufactures Building were sewing machines involved in the production of clothing, as well as a demonstration of saddle-making. Omaha's growing meat packing industry uncovered particular fascination in a large, revolving refrigeration unit. (Photo courtesy of Omaha Public Library.)

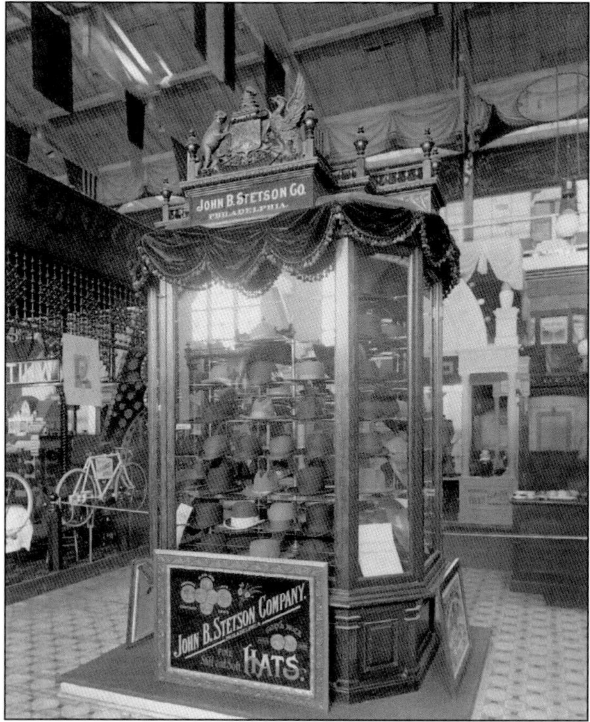

Derbies and other popular styles of hats for men were featured in the display of the John B. Stetson Company of Philadelphia, Pennsylvania. Demonstrating the diversity of products on display in the Manufactures Building, an example of a chainless bicycle is exhibited to the left of the Stetson display case. (Photo courtesy of Omaha Public Library.)

The Omaha Brewing Association received the highest honor awarded at the exposition for their display of draft beer. The exhibits of noted brewers and wine-makers were among the displays of the Manufactures Building. (Photo courtesy of Omaha Public Library.)

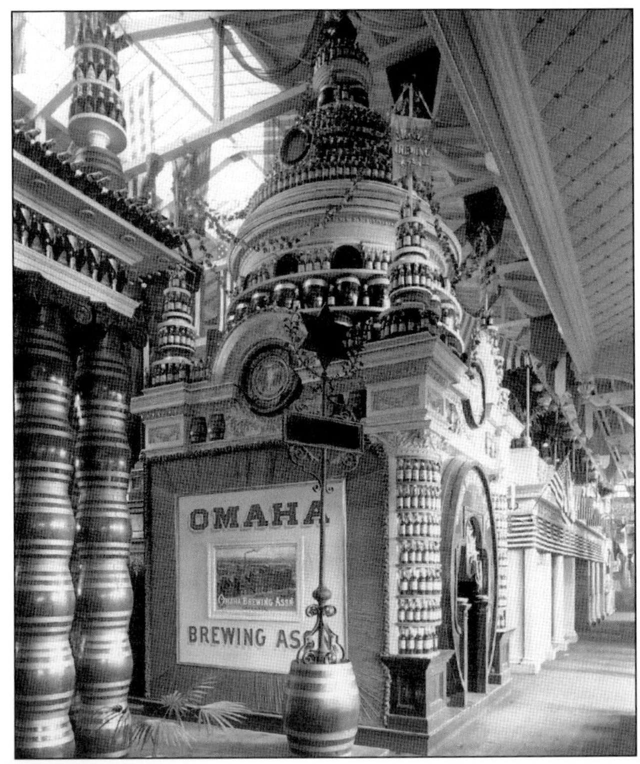

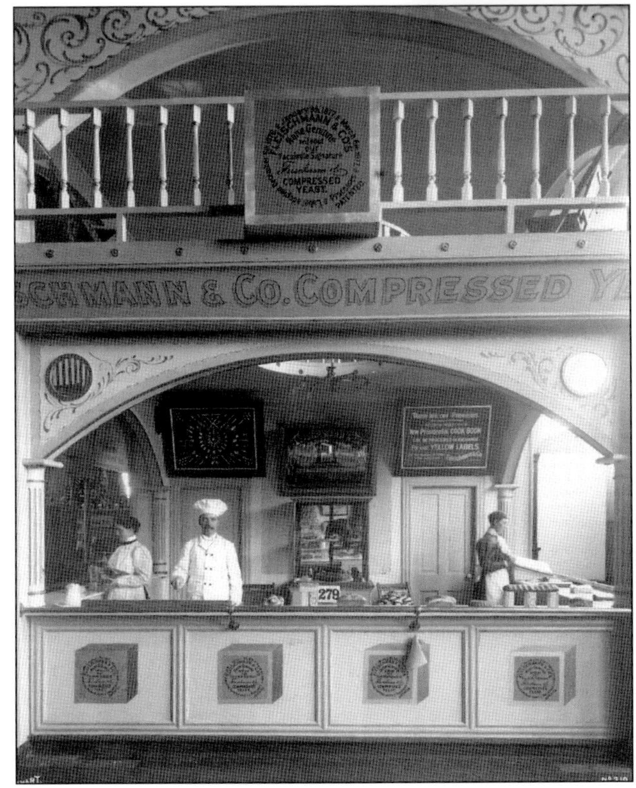

The yeast of Fleischmann and Company was the featured product in this display of bread-making within the Manufactures Building. (Photo courtesy of Omaha Public Library.)

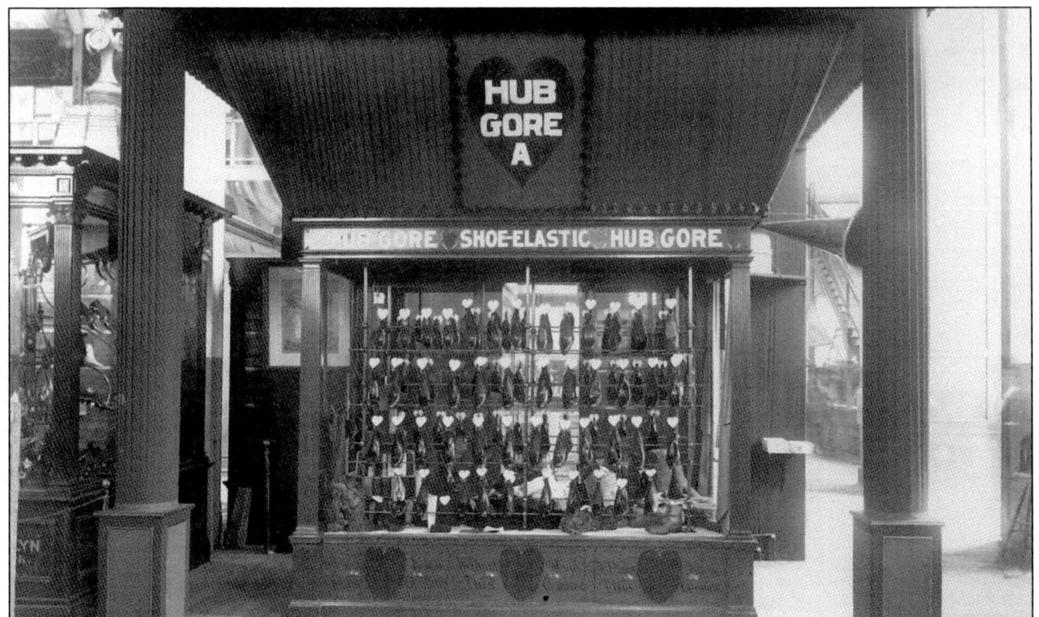

The displays presented in the Manufactures Building included this array of Hub Gore shoes. Featured here are men's and women's shoes and boots. The exhibit booths of the Manufactures Building were among the most elaborate and design-conscious at the exhibition. (Photo courtesy of Omaha Public Library.)

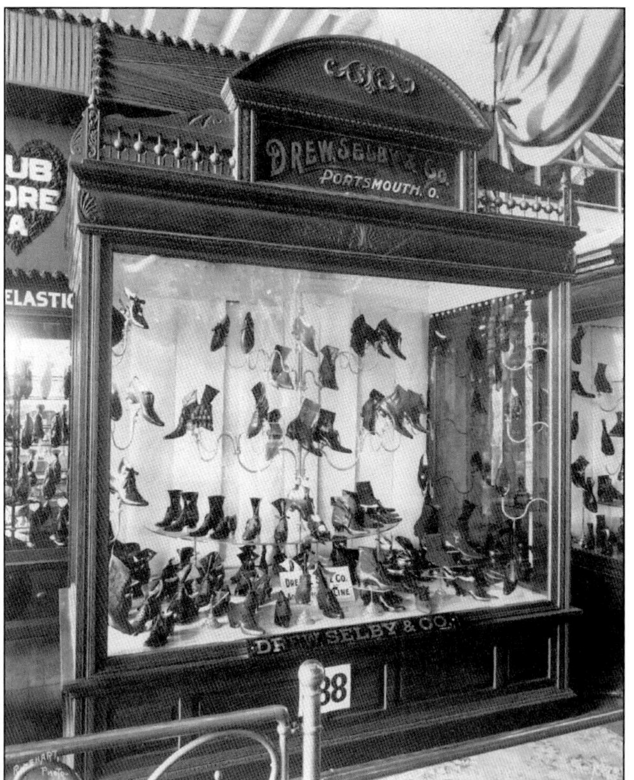

Fine ladies' shoes were featured in this display case by Drew, Selby and Company of Portsmouth, Ohio. The exhibits of other shoe manufacturers were also on display and pictured in this view inside of the Manufactures Building. (Photo courtesy of Omaha Public Library.)

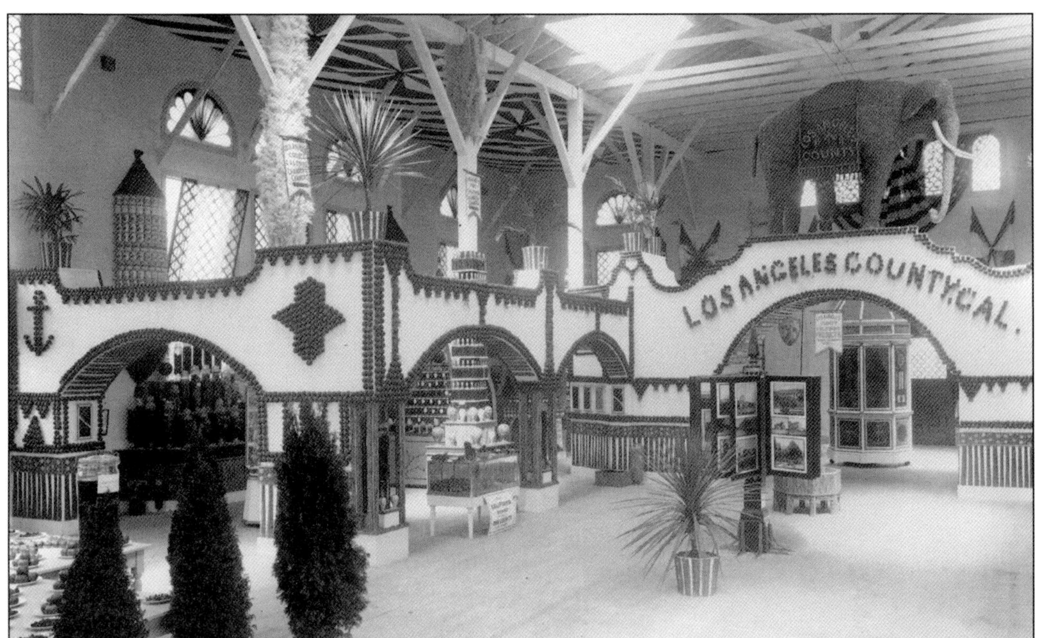

The varied farm products of Los Angeles County, California, were showcased in this display by the Los Angeles County Chamber of Commerce. Natural and processed food products were displayed among examples of native vegetation and photographic representations of California agriculture. The common use of skylights in many Trans-Mississippi Exposition buildings is also evident in this picture. (Photo courtesy of Omaha Public Library.)

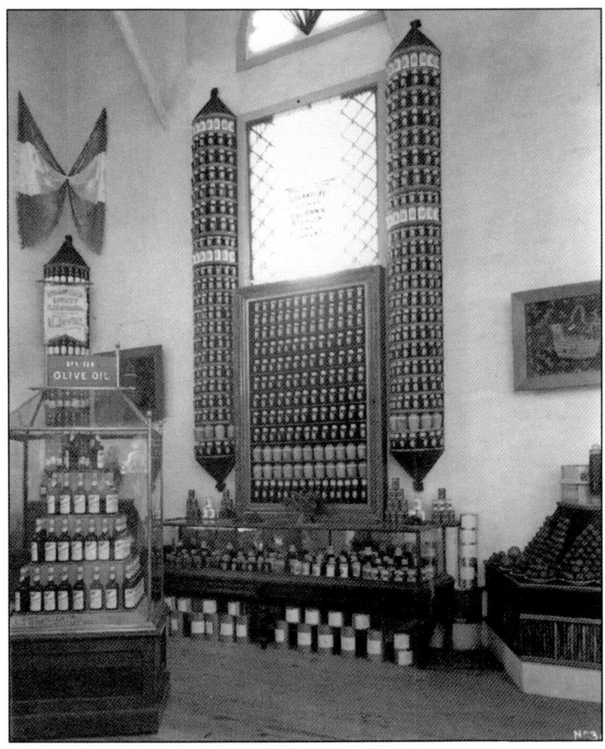

La Crescenta Olive Oil by manufacturer H. Jevne was included in this display from Los Angeles County, California. Beneath and at either side of the large window in this picture is the medal-winning display by Bishop and Company, also of Los Angeles County. Featured here are their preserved, pickled, and crystallized fruit products, including candied citron. (Photo courtesy of Omaha Public Library.)

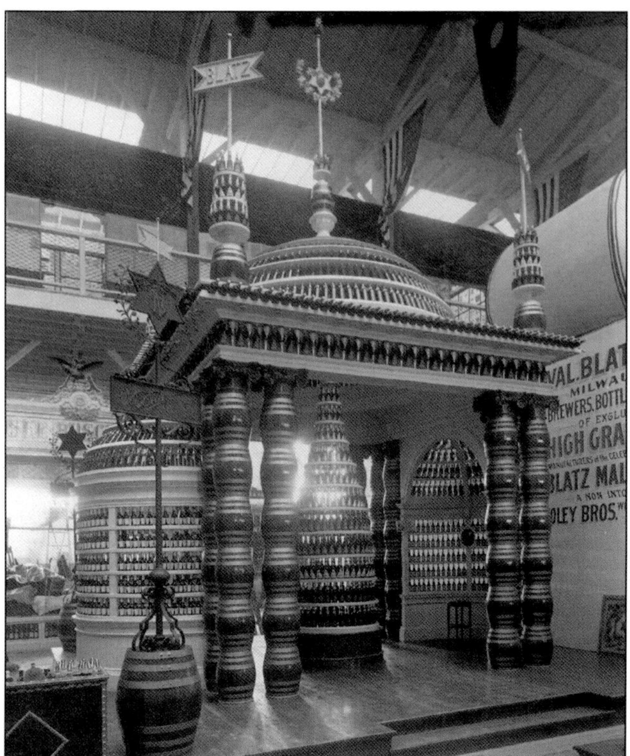

The brewing legacy of Milwaukee's Valentin Blatz (1826–1894) was the centerpiece of this display, constructed from kegs and bottles containing his product. Alcohol at the exposition created a minor controversy as planners considered the question of remaining open for Sunday visitors. It was determined that the exposition would be open from 1:00 p.m. until 10:00 p.m. on Sundays, but no liquor would be sold. (Photo courtesy of Omaha Public Library.)

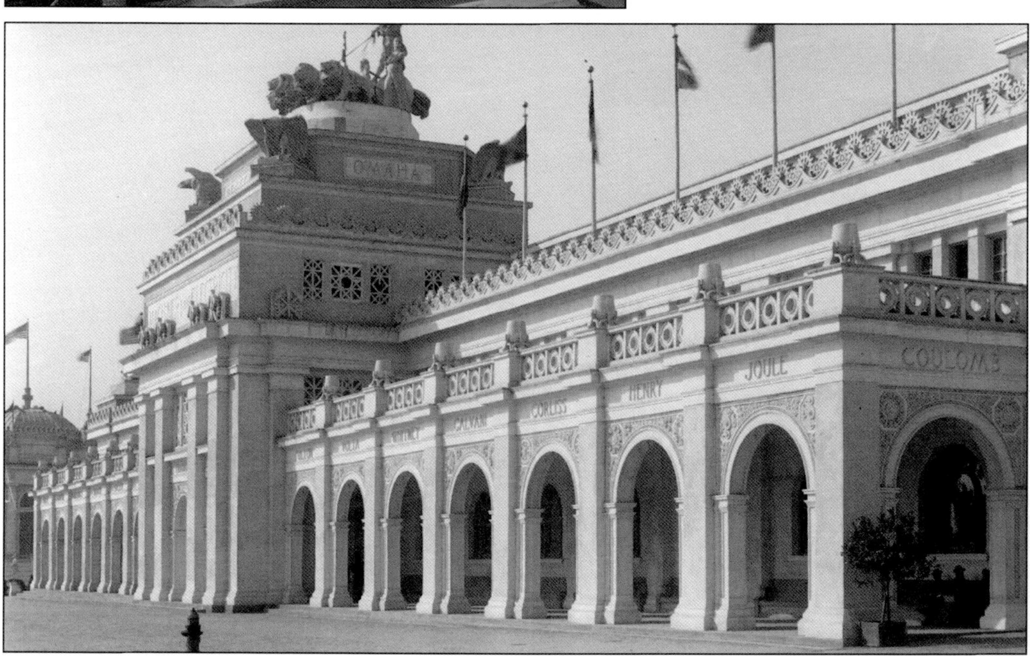

The detail of the Machinery and Electricity Building, as it was seen from the Grand Court, is evident in this photograph. Though no losses to fire occurred on the Grand Court, fire hydrants, such as the one in the foreground, were in place to safeguard this possibility. The loss of the Mt. Nebo Chapel at the Old Plantation display on the North Midway was the most significant fire loss during the exposition season. (Photo courtesy of Omaha Public Library.)

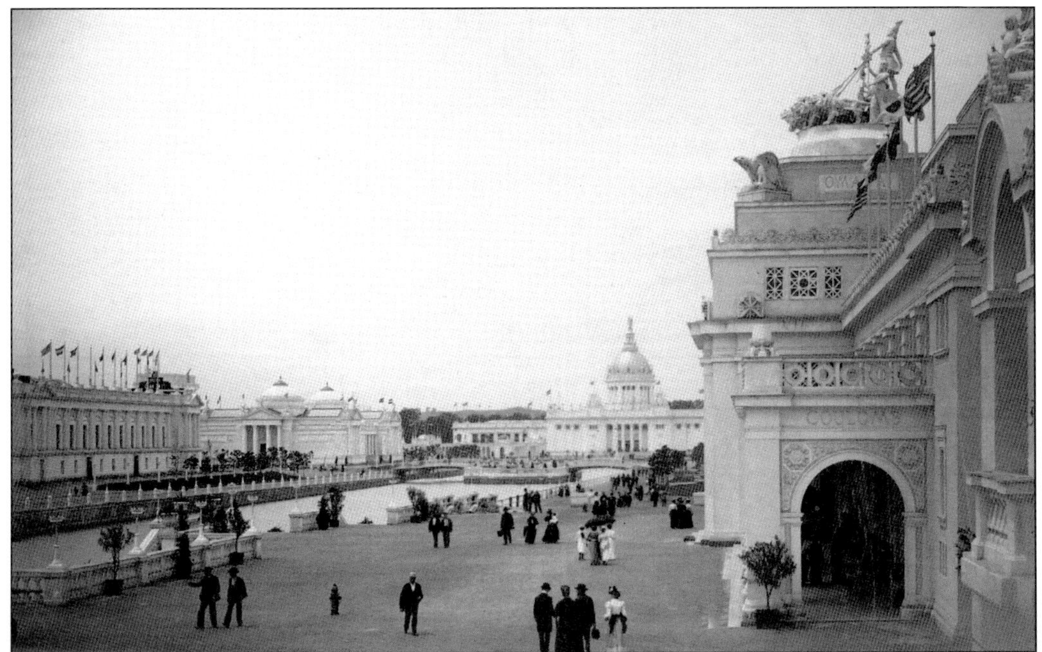
The northeast corner of the Grand Court featured the Machinery and Electricity Building. The ornamentation of this building included a crest of statues and cogwheels designed to represent man's use of machinery to contain and benefit from the forces of nature. Also pictured is the entrance of the portico, fronting the 304-foot facade of the building. (Photo courtesy of Omaha Public Library.)

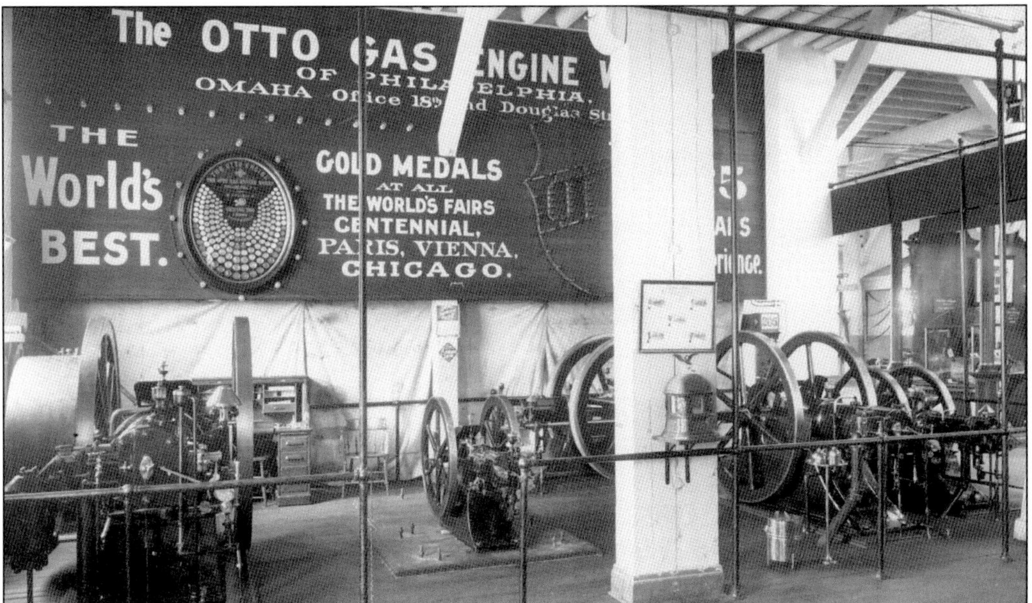

Otto Gas Engines were featured in this display located in the Machinery and Electricity Building. Declaring the status of their product as "world's best," Otto included a display of medals earned at previous expositions, including those honors awarded at the 1893 Chicago World's Fair. (Photo courtesy of Omaha Public Library.)

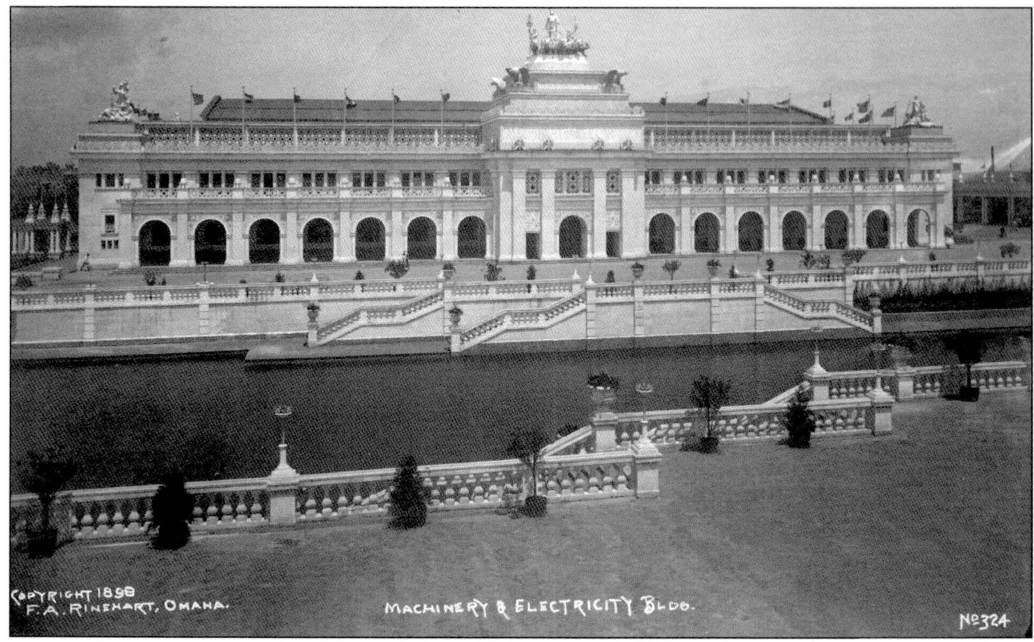

A portico, sixteen feet wide, spanned the front of the Machinery and Electricity Building on either side of the entrance. Skylights and clerestory windows lit displays and offered visitors the opportunity to view scientific wonders of the era, including equipment to supply electricity, telephones, and telegraph equipment. One component of the machinery exhibit was an extensive display of tools. (Photo courtesy of Omaha Public Library.)

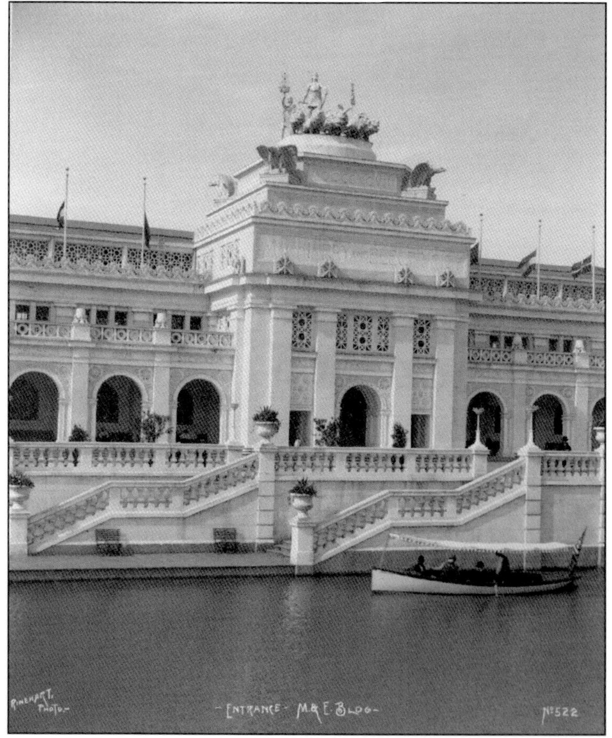

The names of scientists who are now synonymous with innovation, technique, and excellence were featured above the arches of the Machinery and Electricity Building. Among the names were Georg Simon Ohm, James Watt, Benjamin Franklin, and Robert Fulton. The main entrance of this building faced the north side of the lagoon located at the center of the Grand Court. (Photo courtesy of Omaha Public Library.)

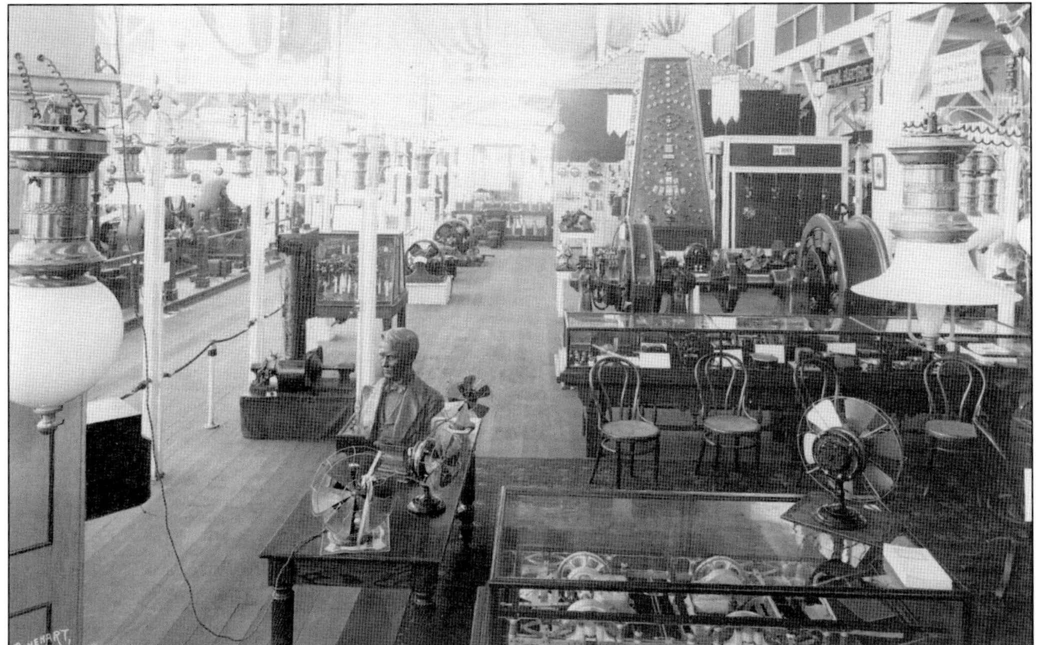

Exhibits of the Machinery and Electricity Building showcased the modern means of electricity production alongside the benefits it provided. Demonstrations of electrical power were presented in a historical perspective that included a range of electricity-generating dynamos, telegraph equipment, and displays illustrating both the history of the telephone and the progress of lighting devices, featuring modern arc lights and incandescent bulbs. (Photo courtesy of Omaha Public Library.)

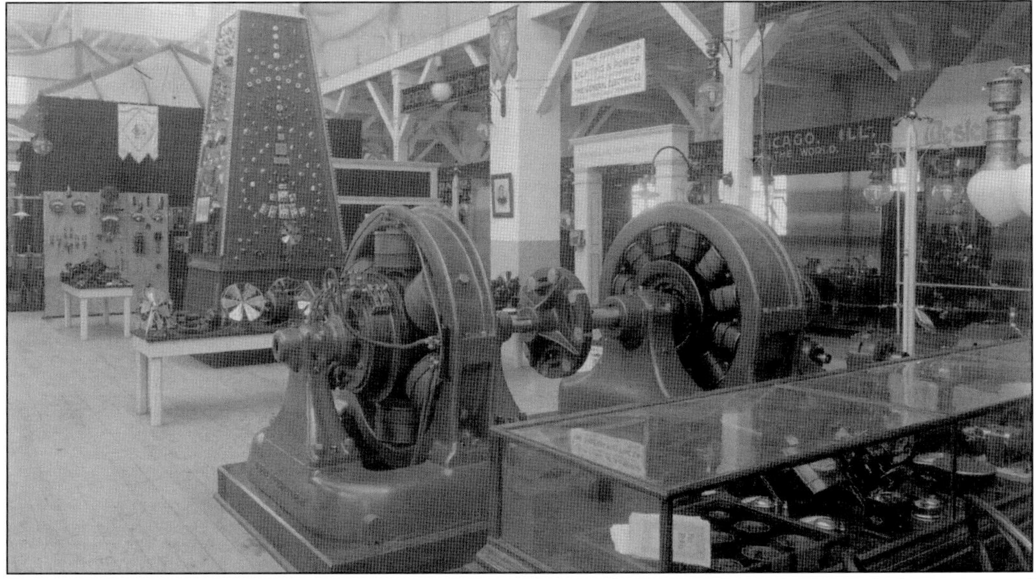

The display of the General Electric Company featured electrical products and the means of generating electricity. The powerhouse near the northeast corner of the exposition grounds housed the boilers, engines, and generators provided by General Electric to supply power to all parts of the exposition. (Photo courtesy of Omaha Public Library.)

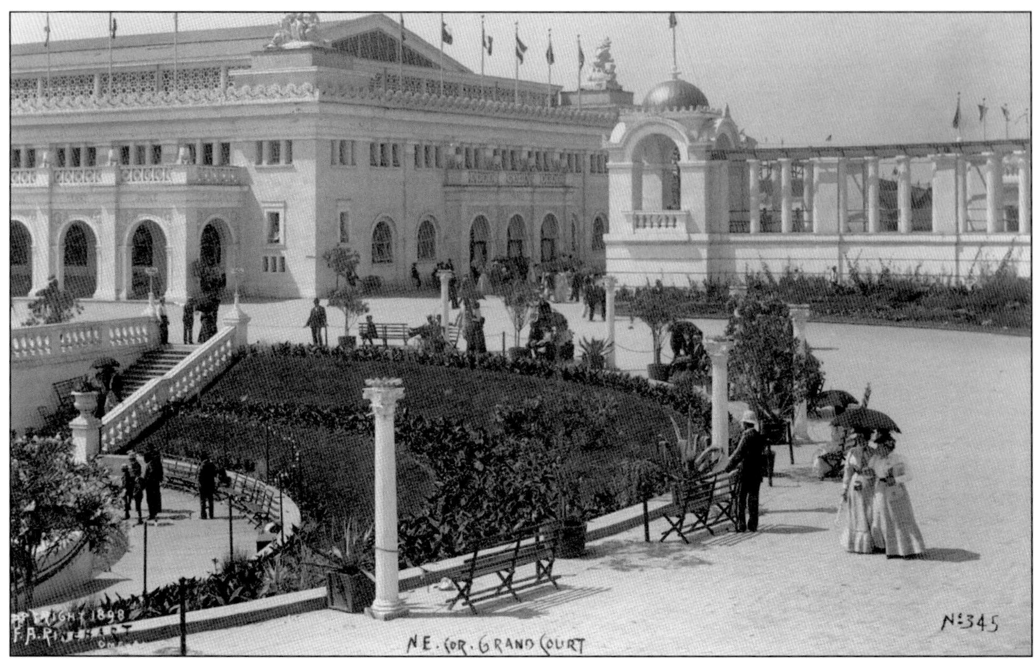

The open space at the east end of the Grand Court, which received considerable landscaping attention, was a popular relaxation area and meeting spot for exhibition visitors. The semi-circular colonnade that divided the Grand Court from the exhibition's South Viaduct approaches the east side of the Machinery and Electricity Building in this view. (Photo courtesy of Omaha Public Library.)

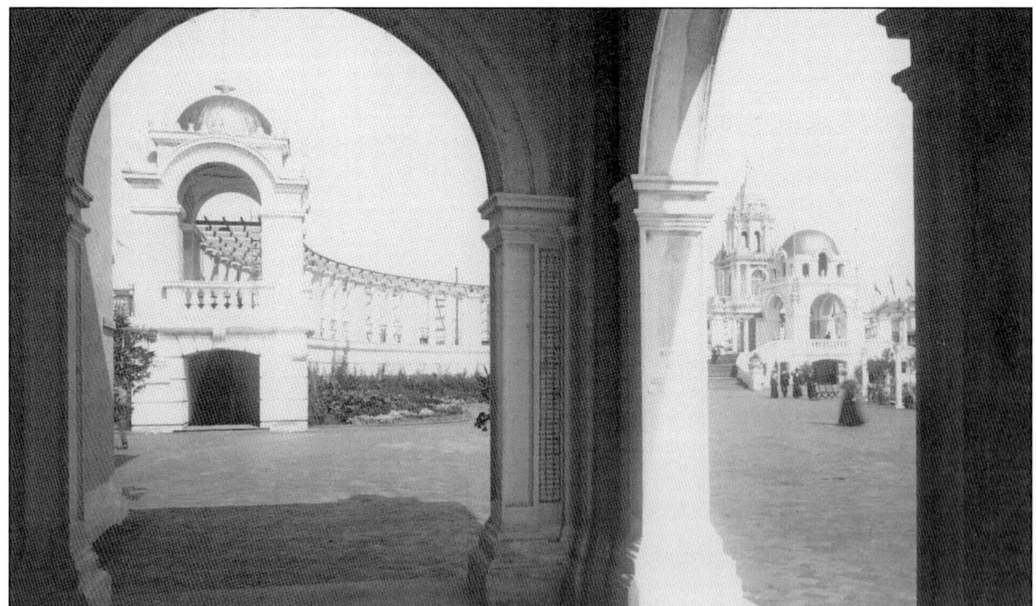

The shade of this 16-foot wide portico, which provided covered passage along the front of the Machinery and Electricity Building, was in stark contrast to the ivory structures of the exposition. The archways pictured here were lined with the small light bulbs that, when lit, defined an evening visit to the exposition for many visitors. (Photo courtesy of Omaha Public Library.)

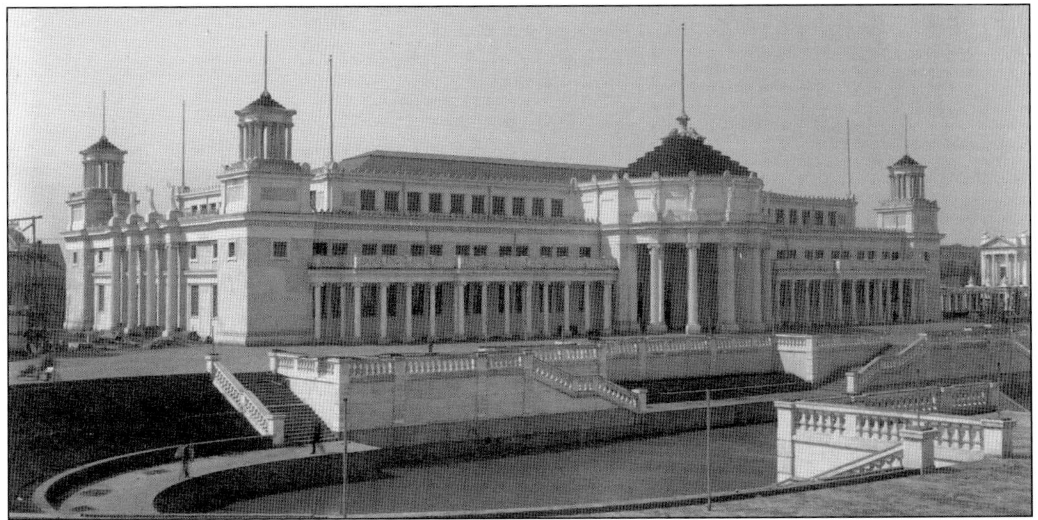

The Mines and Mining Building, located at the southeast corner of the lagoon, featured a central dome, 150 feet in circumference. Coal from a number of key coal-producing states; gold from Alaska, Colorado, and the Black Hills; and South African diamonds were a few among the many displays awaiting visitors interested in natural resources. (Photo courtesy of Omaha Public Library.)

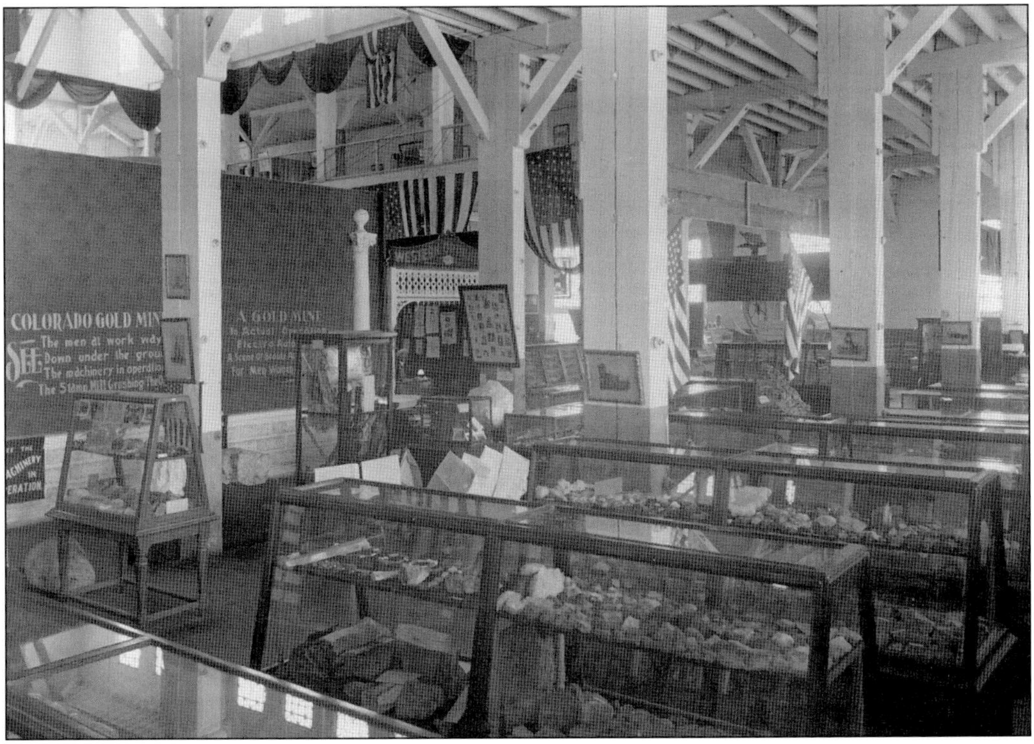

The opportunity to see mining machinery in operation and examples of recently discovered Telluride gold ores were the principal attractions of this mining display from Colorado. Exhibits of rocks and minerals were complemented with explanations of the natural conditions under which they are found and the processes by which they are extracted. (Photo courtesy of Omaha Public Library.)

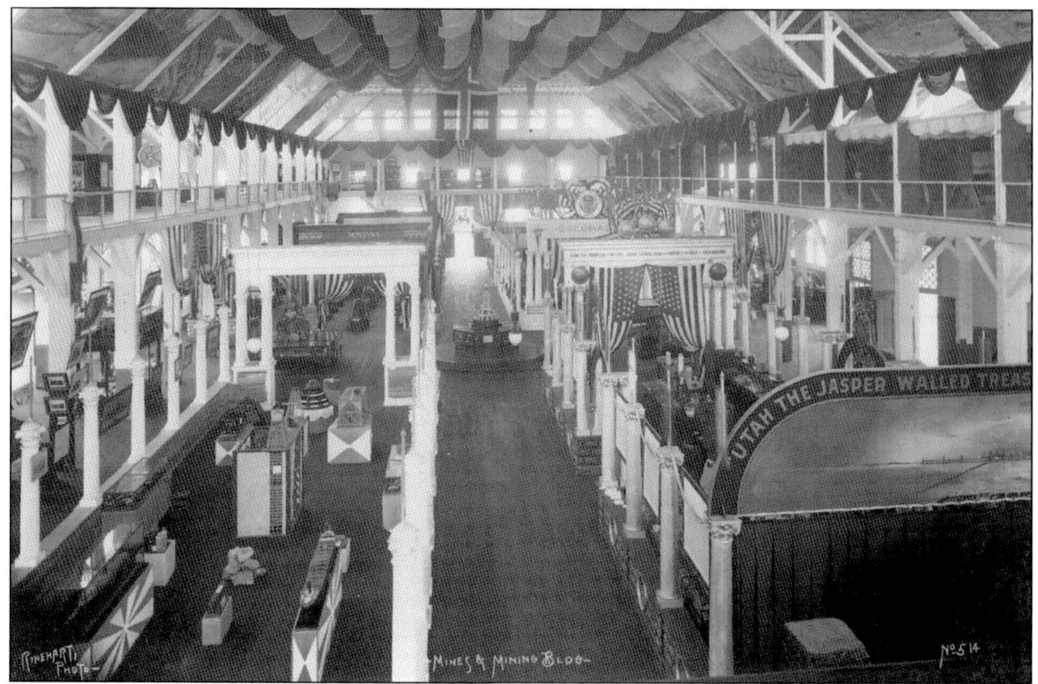

The mineral resources of the Trans-Mississippi region were featured in the displays of the Mines and Mining Building. An effort to display a sample of each mineral produced in this region included the large exhibit, pictured at the right, from Utah. At its entrance, this exhibit boasted, "Utah has produced in 10 years gold, silver, lead and copper in the value of $199,000,000." (Photo courtesy of Omaha Public Library.)

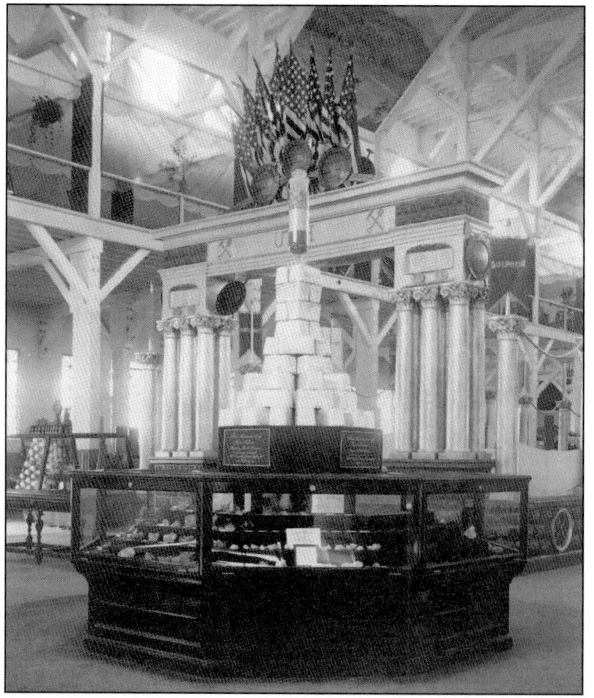

Atop this display case, located at the center of the Mines and Mining Building, are 48 pure silver ingots, weighing a total of 3,488 pounds. This gold-medal-winning exhibit was the product of the Omaha and Grant Smelting Company, founded in 1877 by former Colorado Governor James B. Grant (1883–1885). The entrance to the exhibit of Utah mining products is pictured behind the display case. (Photo courtesy of Omaha Public Library.)

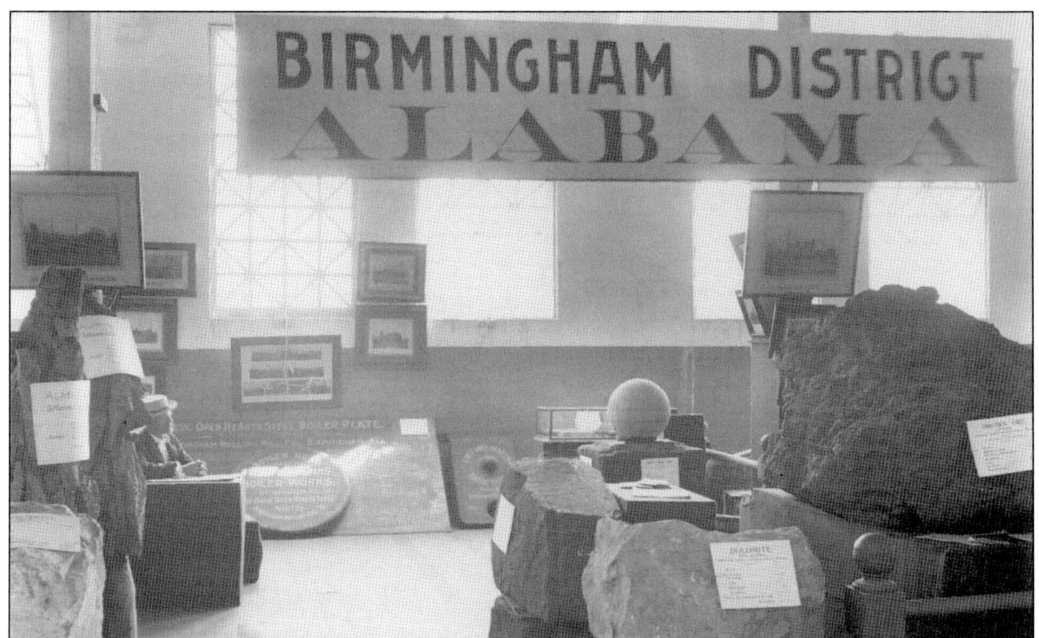

Mining interests near Birmingham, in Jefferson County, Alabama, displayed large samples of the natural resources extracted from their mines. The dark-colored stone at the right is brown ore, while the small light-colored stone at the lower right is dolomite. (Photo courtesy of Omaha Public Library.)

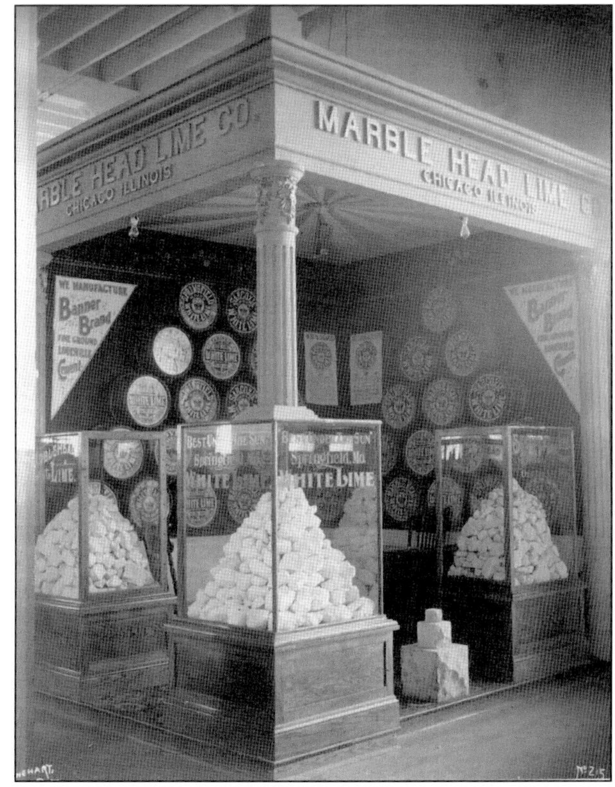

The display of the Marble Head Lime Company of Chicago, Illinois, promoted their products, including white lime from Springfield, Missouri. The reputation of the Springfield product, advertised as the "Best Under The Sun," was upheld with this gold-medal-winning display. (Photo courtesy of Omaha Public Library.)

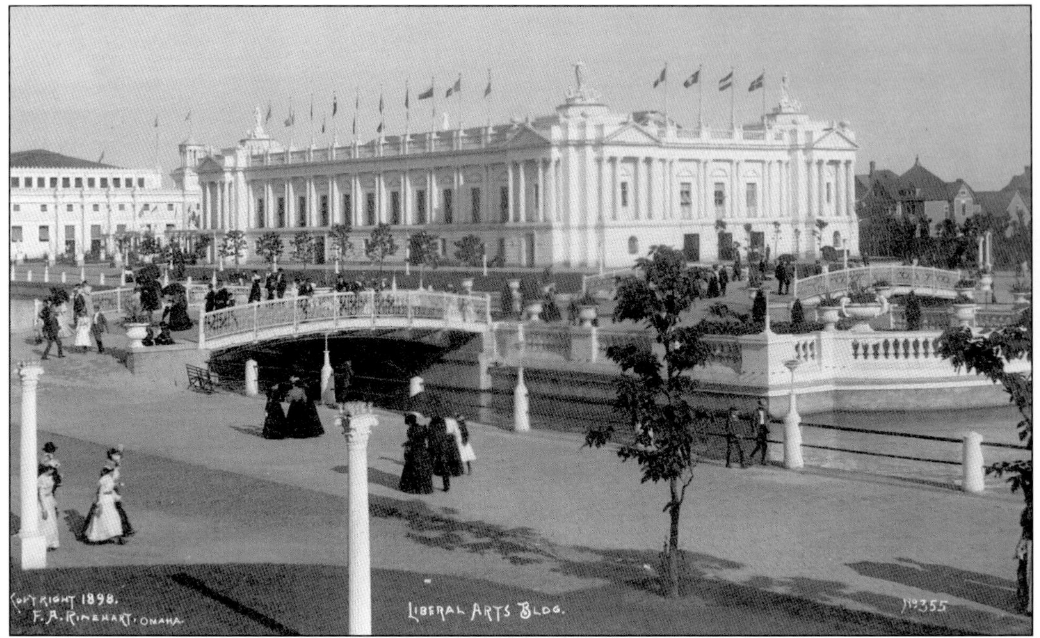

The Liberal Arts Building, a favorite of many exposition visitors, featured modern items familiar to everyone, such as furniture and musical instruments, as well as exhibits promoting advances in chemicals and medicine. A graphophone producing music and a firm exhibiting artificial limbs were among the modern wonders that awaited visitors to this building. (Photo courtesy of Omaha Public Library.)

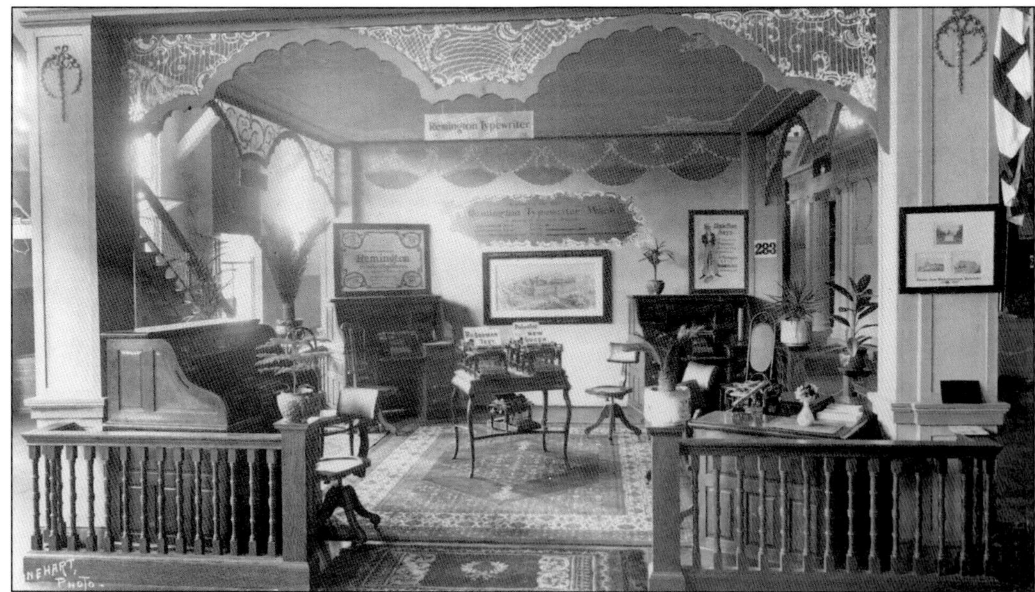

The modern business office was the focus for this exhibit by the Remington Typewriter Works in the Liberal Arts Building. Typewriters designed to produce characters of the German, Greek, and Russian languages were also included. In an age of expositions, a sign on the back wall of this display touted the progress of Remington since the 1876 Philadelphia Centennial Exposition and the 1893 Chicago Columbian Exposition. (Photo courtesy of Omaha Public Library.)

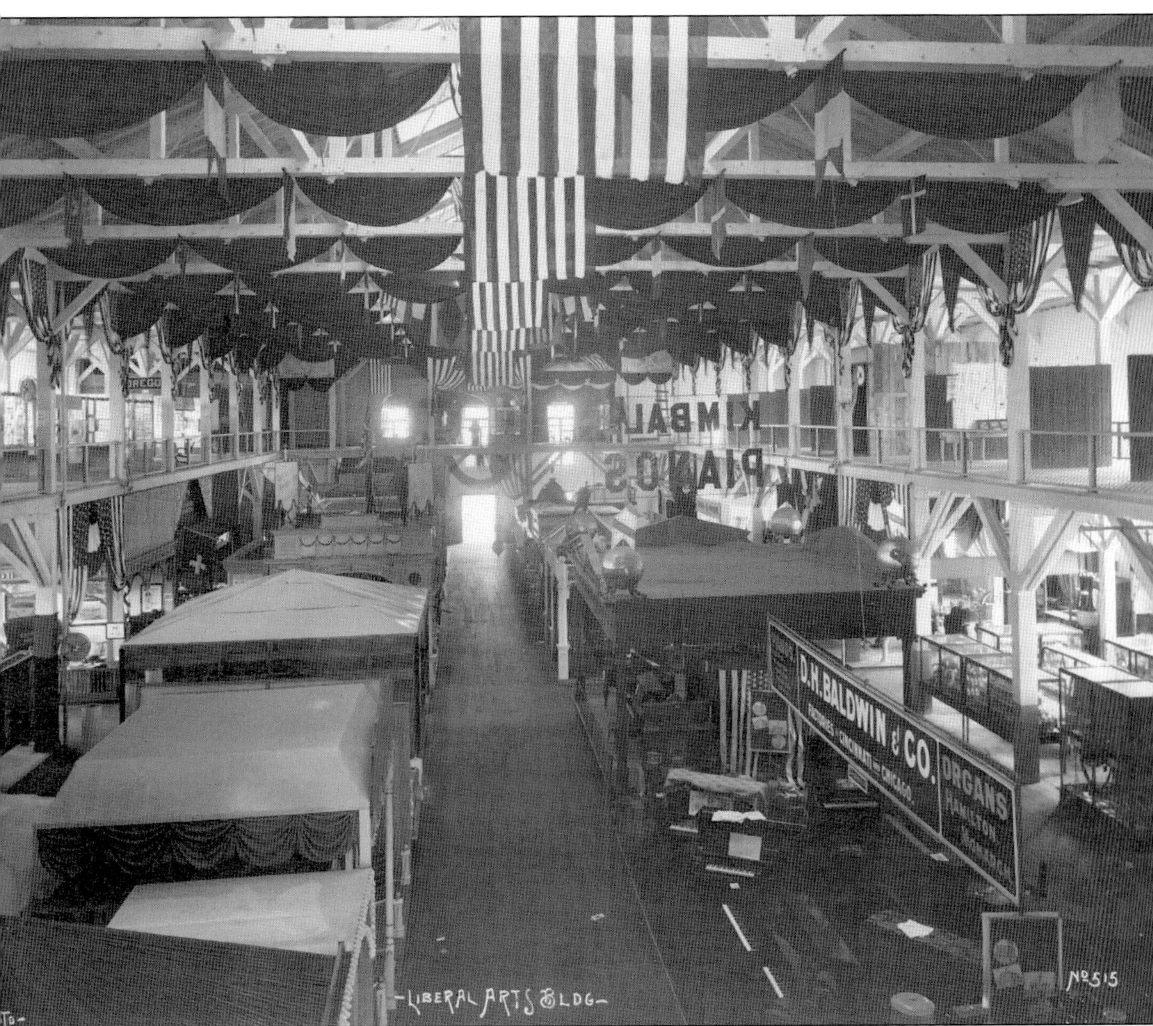

Fabric, furniture, typewriters, jewelry, and medicine were a few among the many varied items exhibited in the Liberal Arts Building. Pictured at the right are the exhibits of Kimball Pianos and Baldwin Organs. Many finer items, such as lace and needlework, were displayed in the galleries above the main floor. (Photo courtesy of Omaha Public Library.)

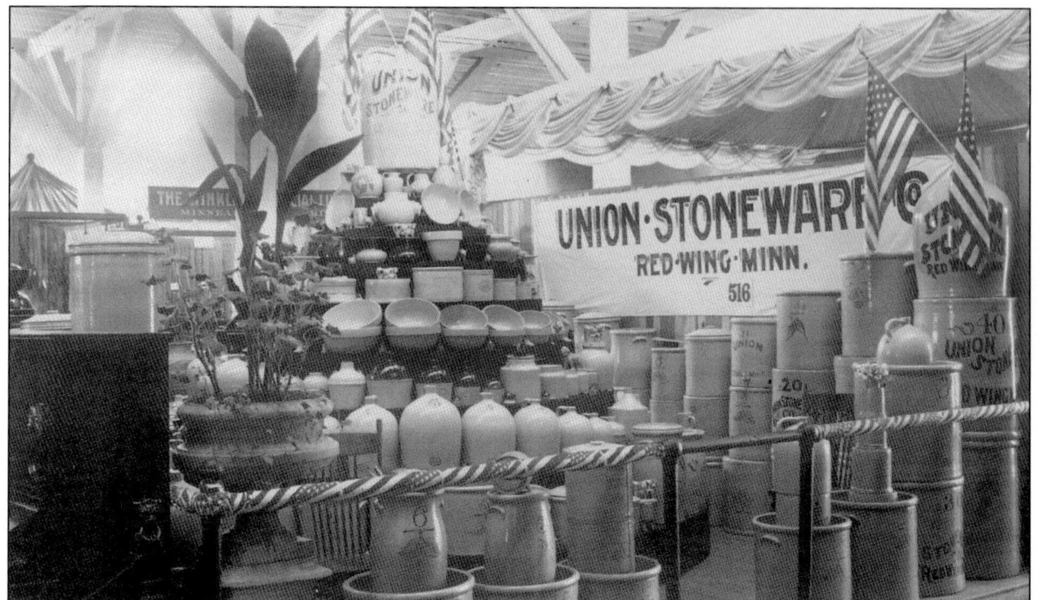

The Union Stoneware Company of Red Wing, Minnesota, producers of items such as jugs, bowls, and crocks, exhibited their products in the Liberal Arts Building. Visitors took the time to view stoneware items ranging from 40-gallon crocks to table-top water pitchers. (Photo courtesy of Omaha Public Library.)

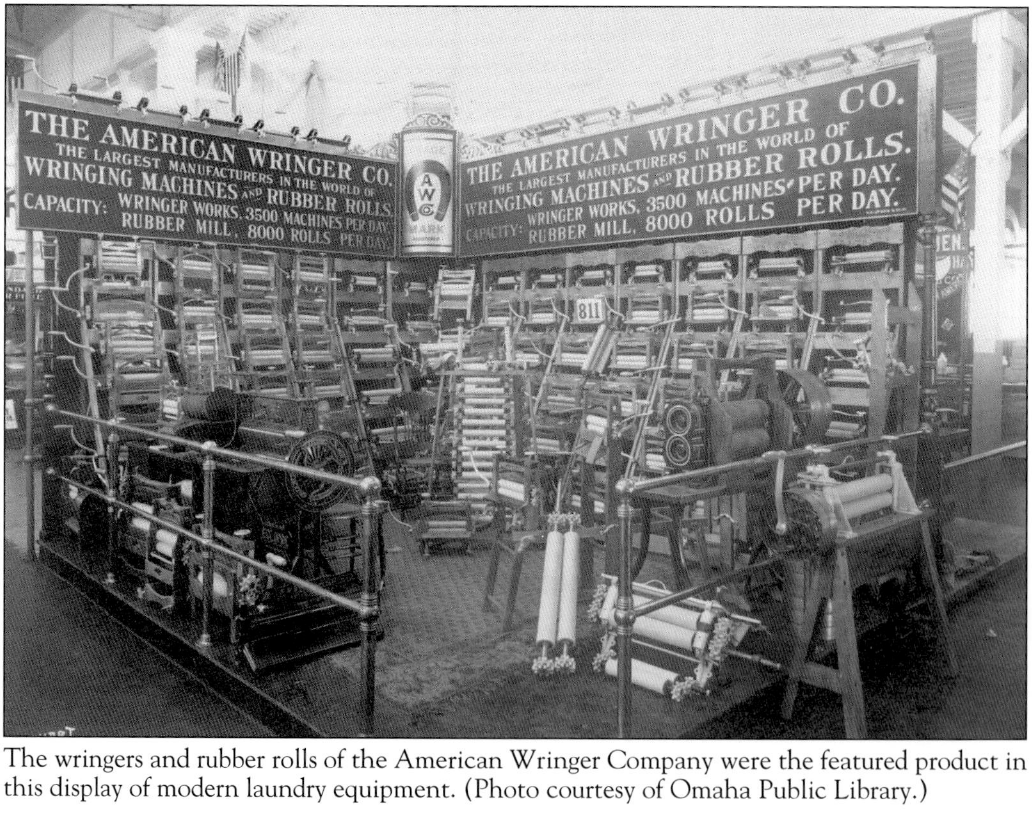

The wringers and rubber rolls of the American Wringer Company were the featured product in this display of modern laundry equipment. (Photo courtesy of Omaha Public Library.)

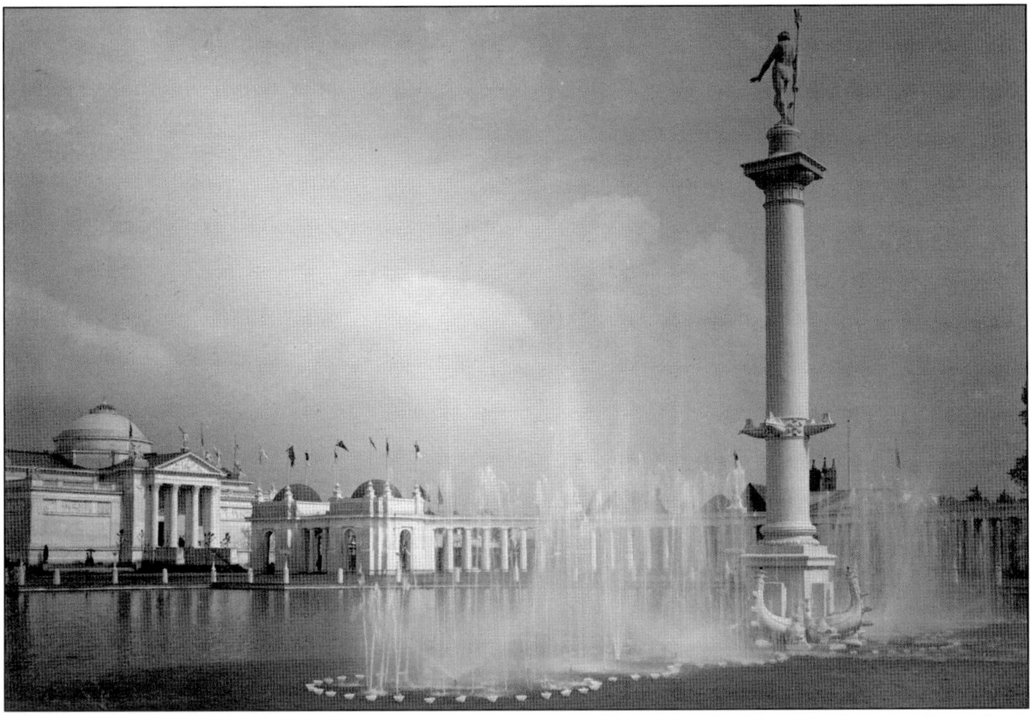

The fountain, designated "Nautilus," at the center of the mirror basin, stood atop a base from which water jets formed a spray. Illuminated in the evening, the sight was equally spectacular during daylight, as water sprayed in an area 100 feet in length and 50 feet in width. (Photo courtesy of Omaha Public Library.)

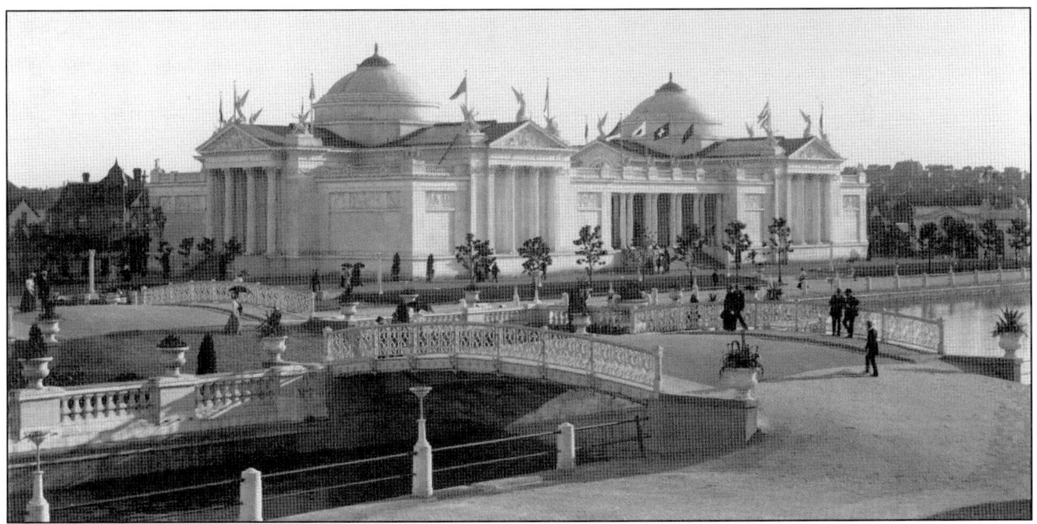

The two domes of the Fine Arts Building were regarded as one of the most attractive features of the Grand Court. This two-building complex was designed to provide for a proper organization and division of the works that were on display. Included in this exhibition was a carbon photograph display of well-known masterpieces. No pictures were hung at a height greater than 10 feet, thus allowing for the appreciation of technique, as well as subject matter. (Photo courtesy of Omaha Public Library.)

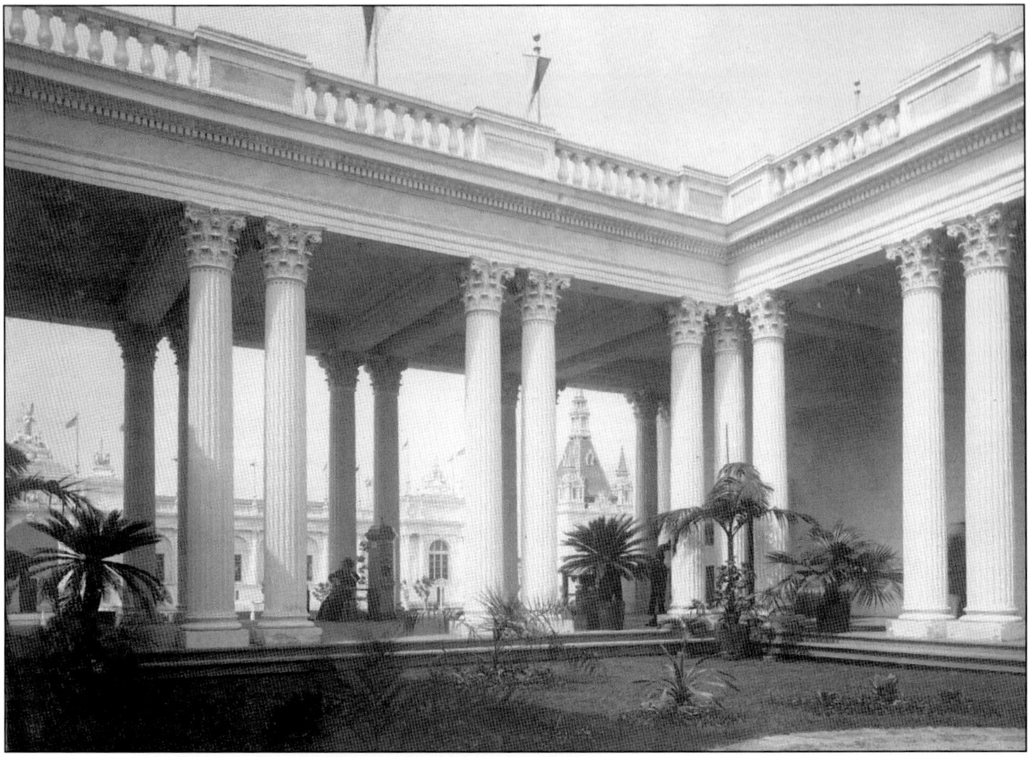

The peristyle, or open courtyard, between the domed structures of the Fine Arts building contained a fountain and flower garden. The Corinthian colonnades provided enclosure for the gardens, as well as providing a setting for exhibits better suited for outdoor display. (Photo courtesy of Omaha Public Library.)

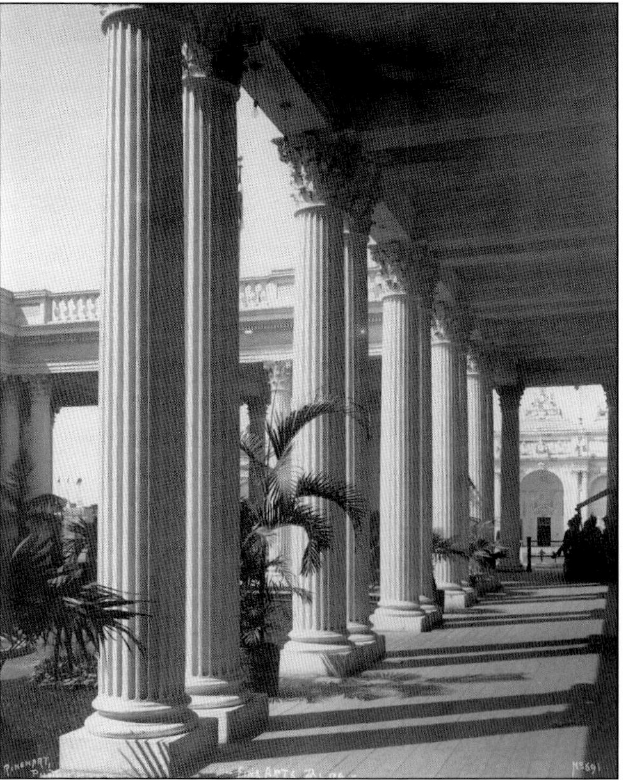

The open courtyard between the two structures comprising the Fine Arts Building was a peaceful retreat for exposition visitors. Passage between the two structures was available via the paved walkway or paths leading visitors through the garden. (Photo courtesy of Omaha Public Library.)

Pictured with one of two domes atop the Fine Arts Building is the rooftop detailing featured on its exterior and a statue, a likeness of which was placed above each entrance. (Photo courtesy of Omaha Public Library.)

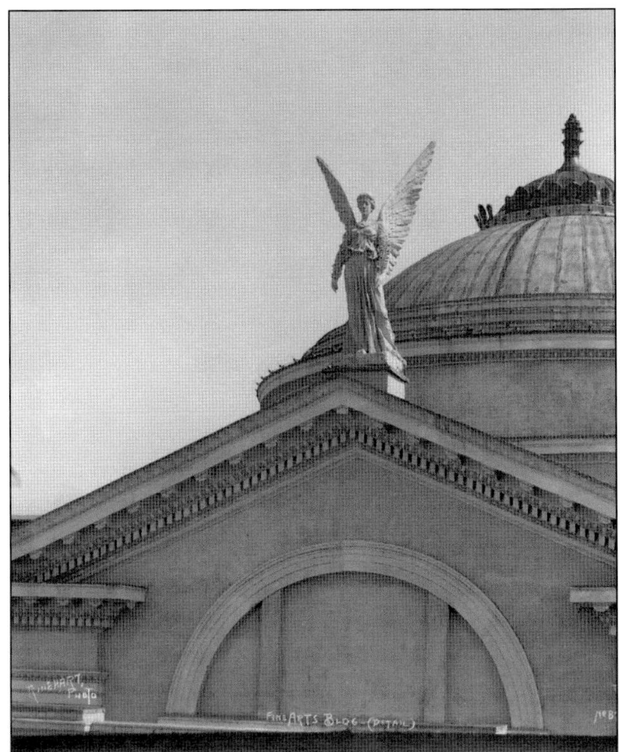

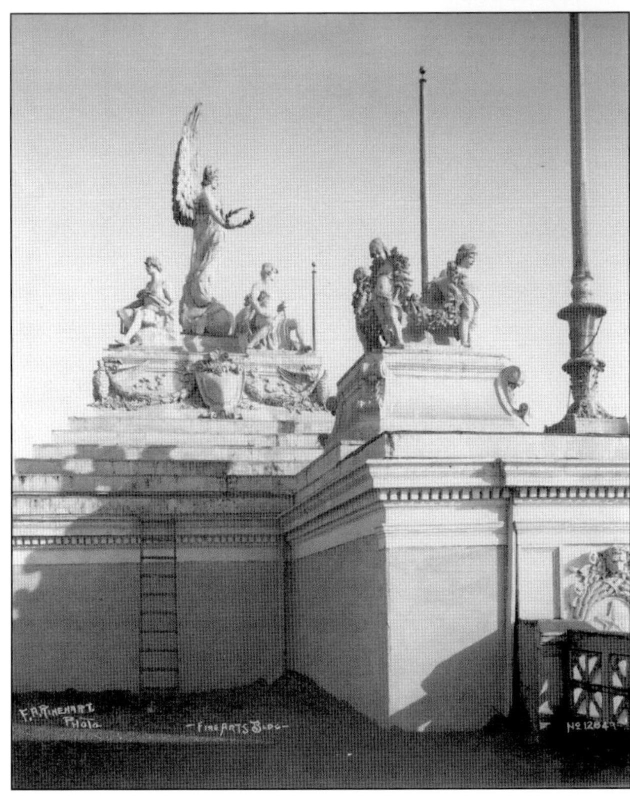

Entitled "Fame," a winged figure of a woman holding a palm branch in her hand is accompanied by statuary atop the Fine Arts Building. This ornamentation was typical of the decorative architectural elements featured on the Grand Court. Exposition organizers sought to limit excessive ornamentation, while still allowing architects to distinguish their work. (Photo courtesy of Omaha Public Library.)

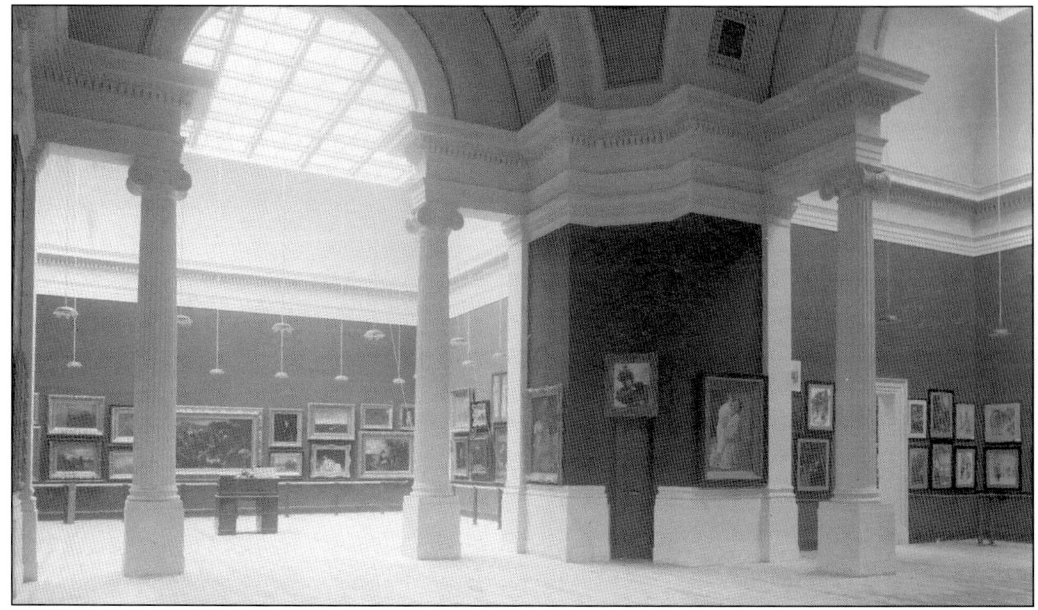

The display of art in the Fine Arts Building included approximately 700 paintings, as well as many statues. While many works were reproductions of noted masterpieces, among the significant original paintings on display was "Return of Spring" by William Adolphe Bouguereau. This 1886 painting, depicting a young woman in the nude surrounded by cupids, was damaged in 1890 and again in 1976 by attackers who did not appreciate its theme. The painting is currently on permanent display in Omaha's Joslyn Art Museum. (Photo courtesy of Omaha Public Library.)

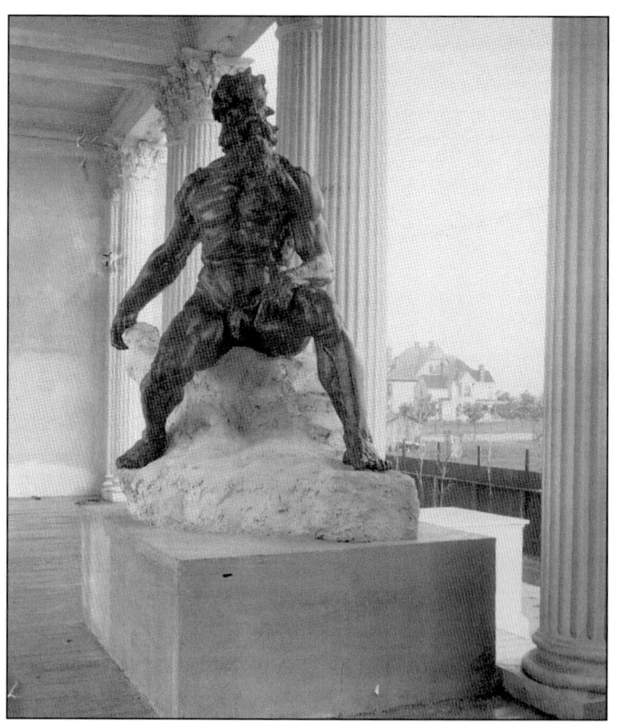

This statue of Neptune stood watch along the south wall of the Fine Arts Building. Pictured here is the detailed work of the Corinthian columns, featured on all four sides of the building. The home in the background is along Twentieth Street, which provided access to the exposition through the Arch of the States. (Photo courtesy of Omaha Public Library.)

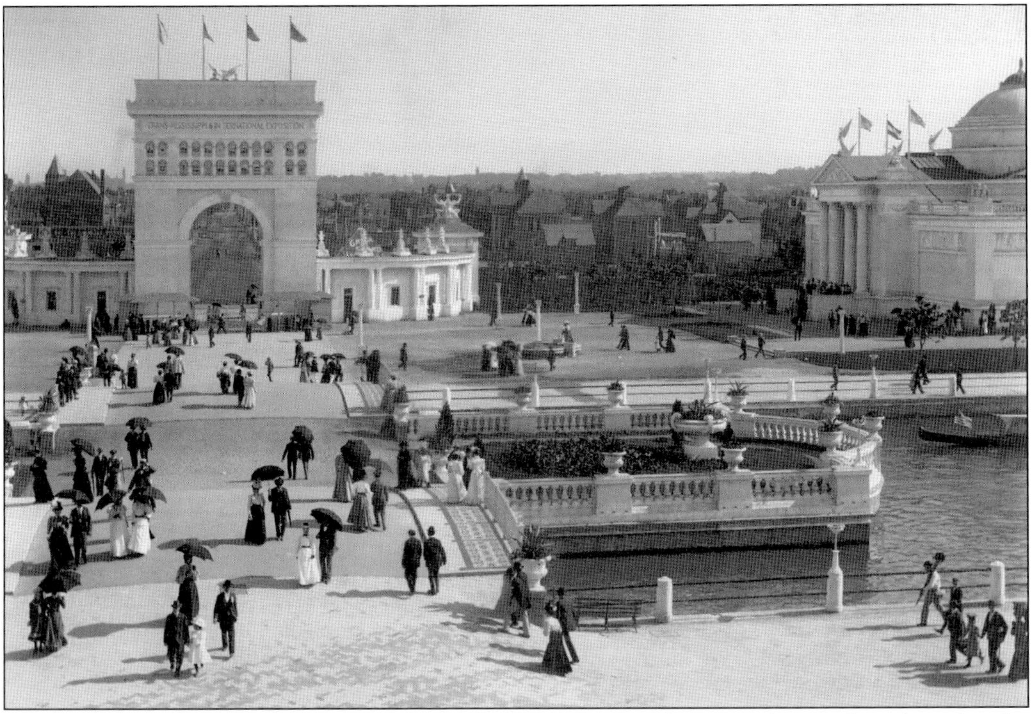

The bridge at the center of the lagoon allowed visitors to pass between the north and south sides of the Grand Court. Seen from atop the Administration Arch is Twentieth Street, pictured through the passageway in the Arch of the States. (Photo courtesy of Omaha Public Library.)

The cornerstone for the Arch of the States was laid on Arbor Day, April 22, 1897. The arch, located at Twentieth and Pinkney Streets, served as the main entrance of the exposition. The 68-foot-high structure was originally planned to be constructed of stone from each of the Trans-Mississippi states and territories and remain as a permanent fixture in Omaha's Kountze Park. While this plan did not come to fruition, the celebration surrounding the Arbor Day dedication event was observed as a holiday by many Omaha businesses, including the significant railroad and warehousing industries. (Photo courtesy of Omaha Public Library.)

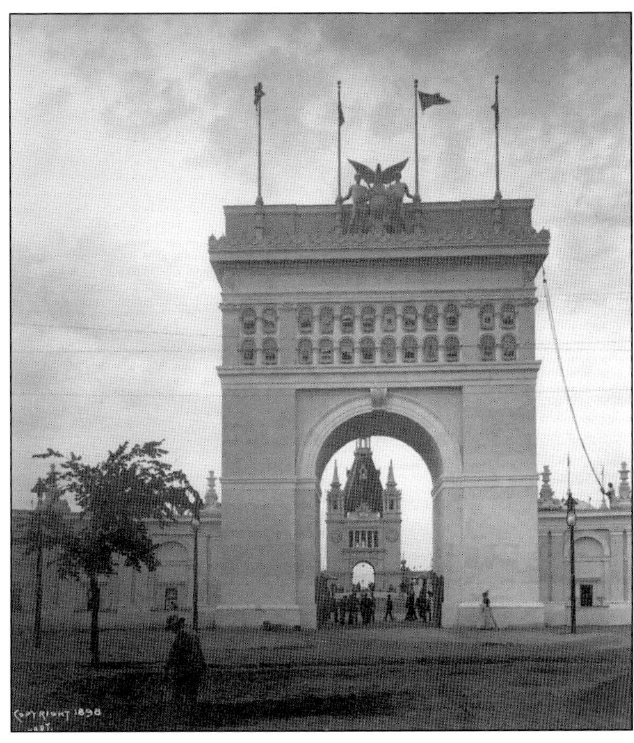

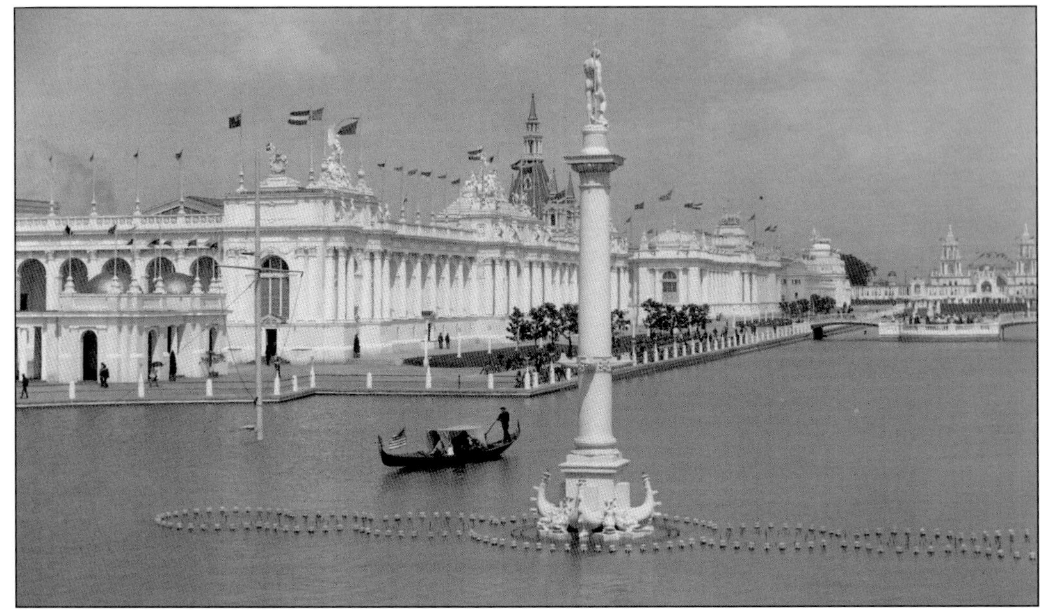

The jets that produced the spectacular water display at the center of the mirror basin surround the "Nautilus" fountain. Pictured at the right, in the distance, are the arched pedestrian bridges at the center of the lagoon. (Photo courtesy of Omaha Public Library.)

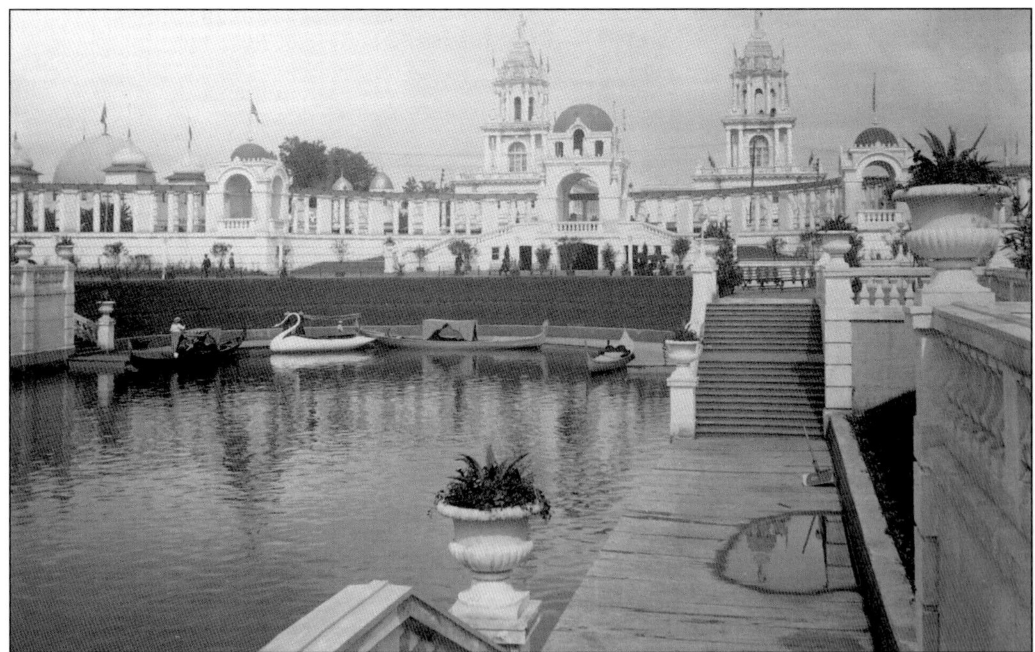

The towers of the twin restaurants provide the backdrop for this view of the lagoon and colonnades at the east end of the Grand Court. The South Viaduct, as the passageway beyond the colonnades was known, crossed Sherman Avenue. Visitors heading for the East Midway or the state buildings of the Bluffs tract used this route. (Photo courtesy of Omaha Public Library.)

Crossing Sherman Avenue at the east end of the Grand Court was the South Viaduct. For a fare of 5¢, exposition visitors could reach the stop at this location along the Omaha Street Railway. Many visitors used this popular line to reach the exposition from downtown Omaha. (Photo courtesy of Omaha Public Library.)

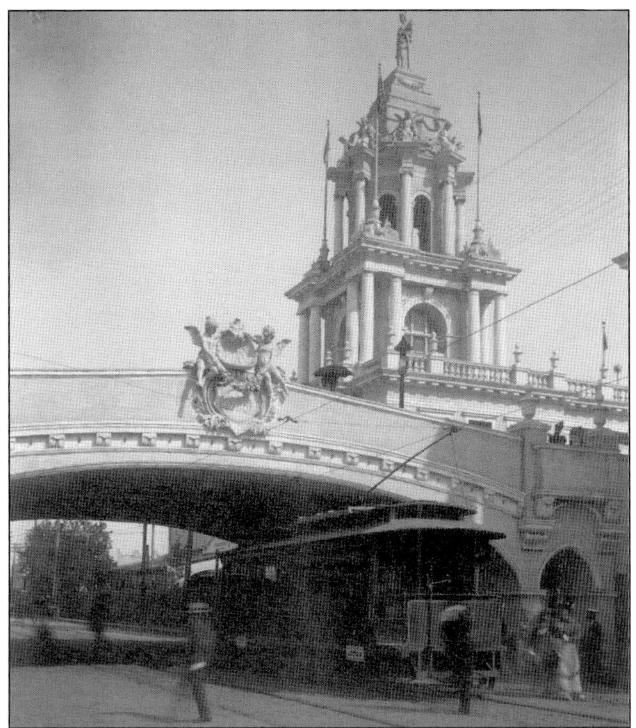

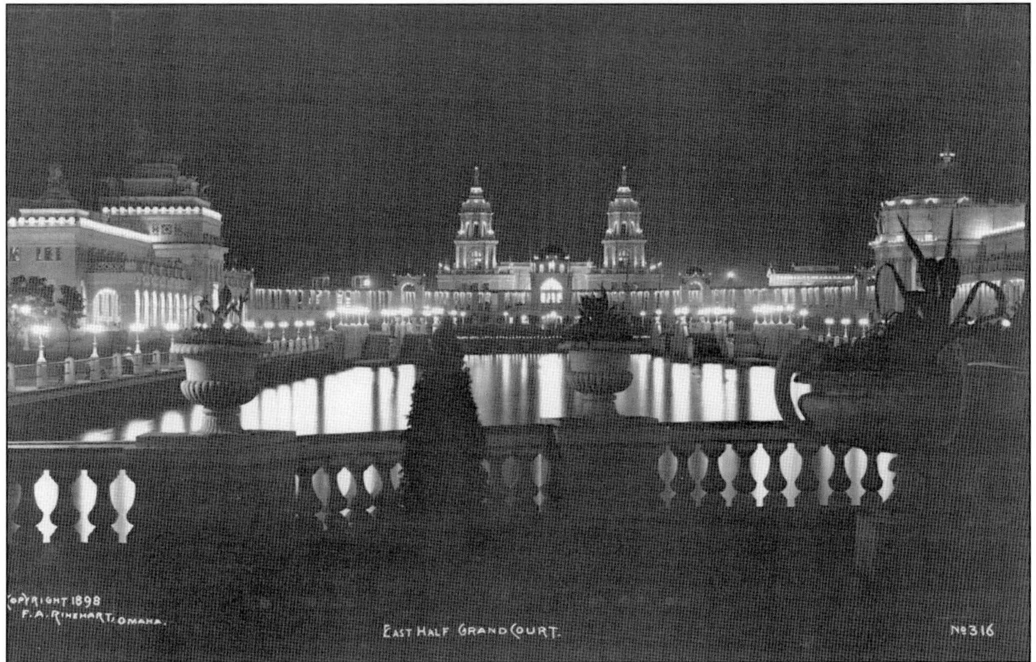

The bridge at the center of the lagoon offered nighttime visitors this view of the two towers located at the South Viaduct of the exposition. Also casting reflections on the still lagoon are the Machinery and Electricity Building (left) and the Mines and Mining Building (right). Exposition visitors enjoyed their evenings during strolls along the Grand Court or from benches positioned along its walking path. (Photo courtesy of Omaha Public Library.)

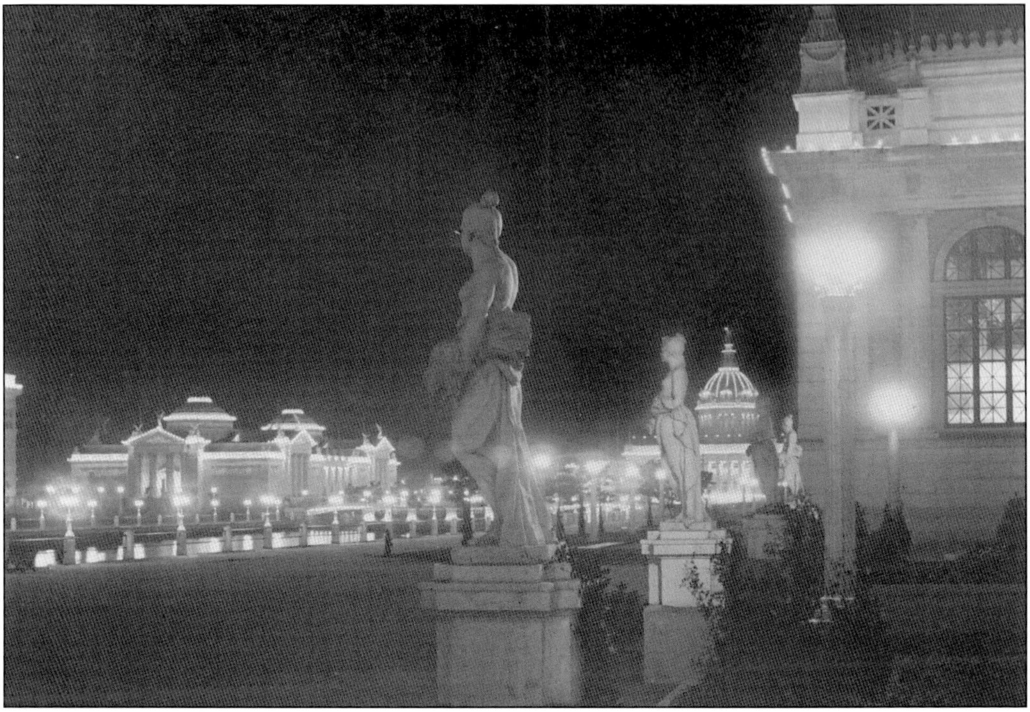

The lighted peaks, domes, and balconies of Trans-Mississippi structures are featured in this view from the north side of the Grand Court near the Manufactures Building. Atop the distant dome is a single light representing a torch, triumphantly hoisted by Bartholdi's "Liberty Enlightening the World." (Photo courtesy of Omaha Public Library.)

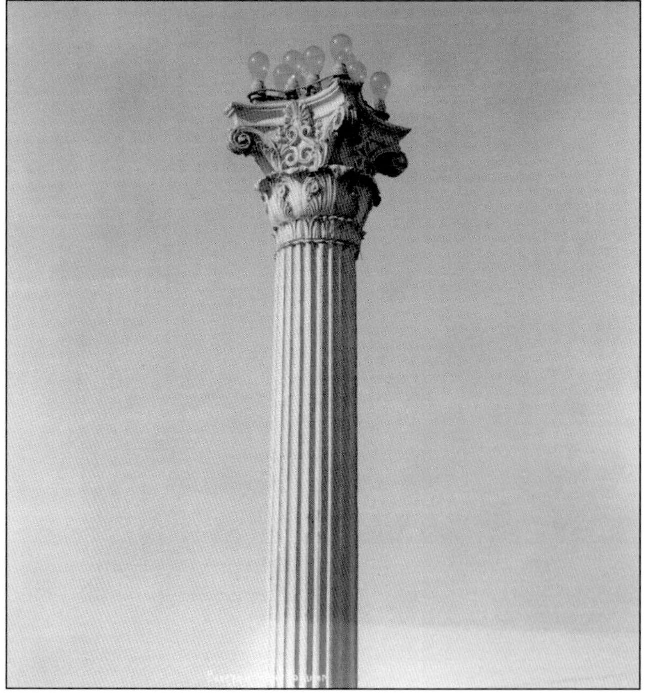

Every evening, the Grand Court of the Trans-Mississippi Exposition was lit by bulbs atop 309 lamp posts, each approximately 10 feet high. The lamp posts contained 12 to 20 bulbs, each producing a soft lighting effect of 8- or 16-candle power. (Photo courtesy of Omaha Public Library.)

Three

ON WITH THE DANCE

A walk along the midway of the Trans-Mississippi Exposition provided visitors with an experience contrasting with the Grand Court and other exhibit-driven areas. The air was rich with the sights, sounds, and smells of a nation overdue for emergence from a decade of challenges.

For less than one dollar, an entire family could enjoy one of the many exhibits, performances or other curiosities available on the busy midway. The familiar Cyclorama presented visitors with a grand view of history unlike any picture book. The Haunted Swing and Giant See-Saw challenged riders with differing perspectives of the world around them. The Streets of All Nations along with the German and Chinese Villages introduced an international flavor, literally and figuratively speaking, to the day of a typical exposition visitor.

Visitors rushed to the east, north, and west stretches of the midway for occasional parades and other celebrations. During his visit, President William McKinley himself saw it fit to walk along the midway, greeting visitors along the way, as he returned to the Markel Café for dinner on the occasion of President's Day at the exposition.

Expressing an ever-present awareness of maintaining respectability, the final report of the "Concessions and Privileges Department," charged with the task of overseeing the Trans-Mississippi Exposition midway attractions, recorded their initial concerns regarding visitors who would follow the guidance of Mark Twain, whose self-proclaimed motto was "On with the dance, let joy be unconfined." The echoes of history remember a "dance," without which the Trans-Mississippi Exposition would remain silent.

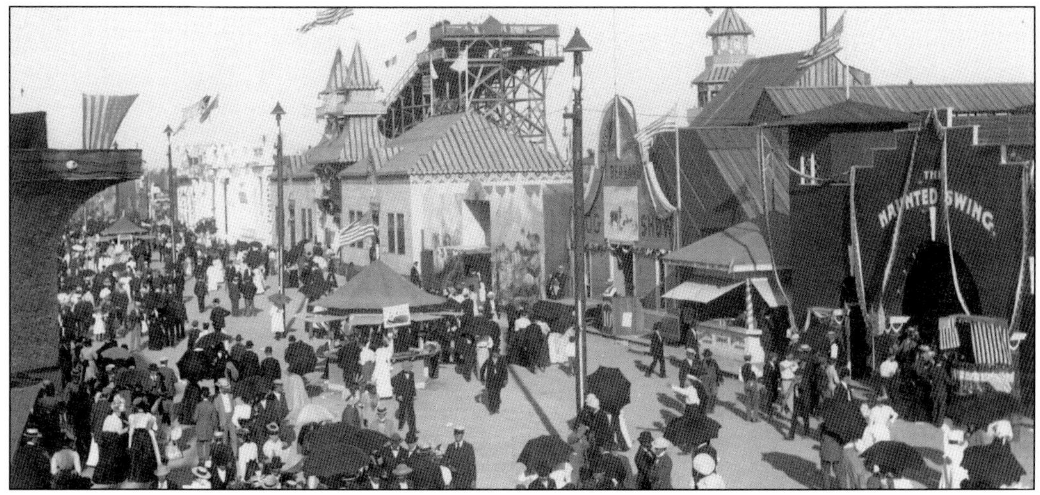

The excitement of the exposition midway attracted many visitors. The incline of the "Shooting the Chutes" thrill ride is seen at the center of this picture. The dark-colored buildings at the right are those that housed a dog show and the "Haunted Swing" attraction. While seeking an opportunity for diversion from the more thoughtful elements of the exposition, directors attempted to preserve the respectability of midway entertainment. (Photo courtesy of Omaha Public Library.)

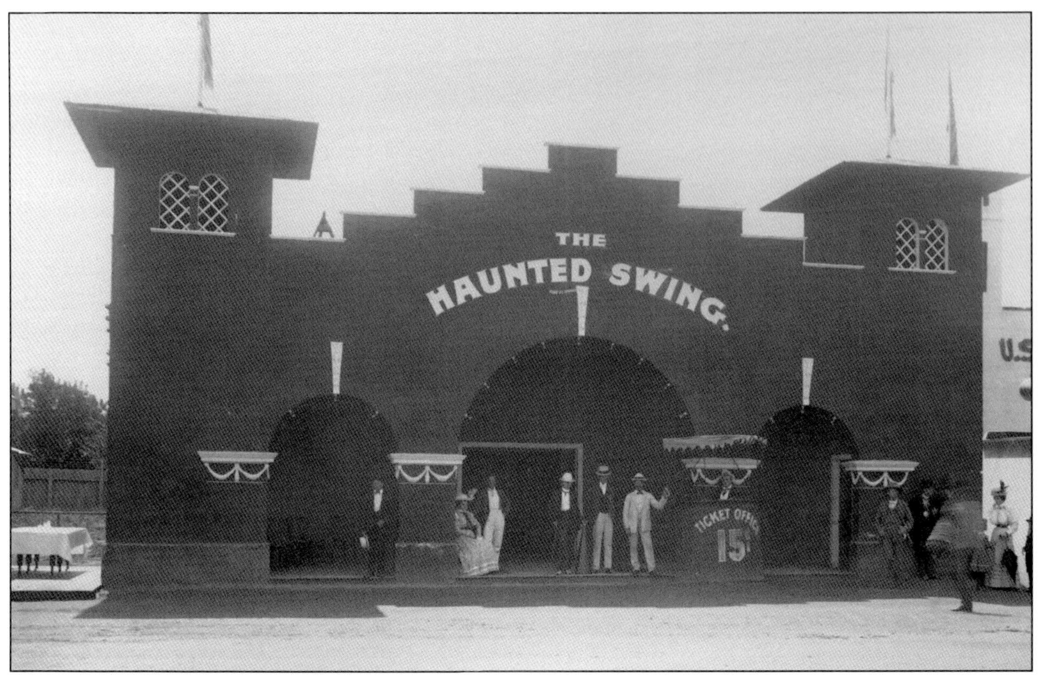

The Haunted Swing was a popular carnival attraction at the time of the exposition. Visitors, seated on a swing in the center of the main room, experienced the illusion of being inverted. The room appeared to turn upside down for those who rode the swing, described by the official exposition guide book as something that must be seen to be enjoyed. (Photo courtesy of Omaha Public Library.)

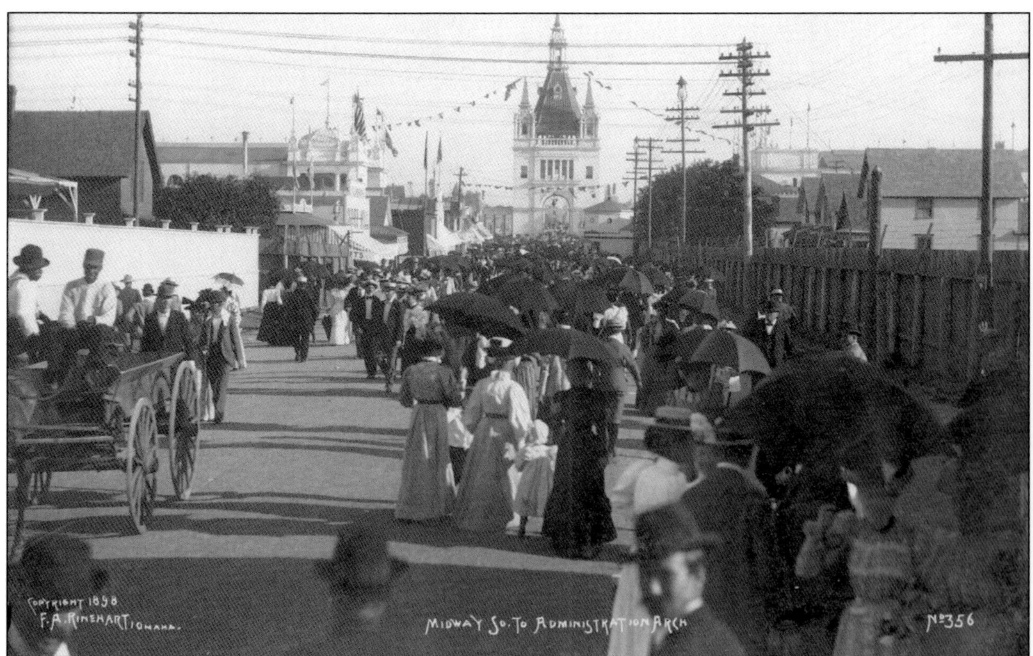

The Administration Arch, located on the north side of the Grand Court, sits at the end of this view of the busy West Midway. (Photo courtesy of Omaha Public Library.)

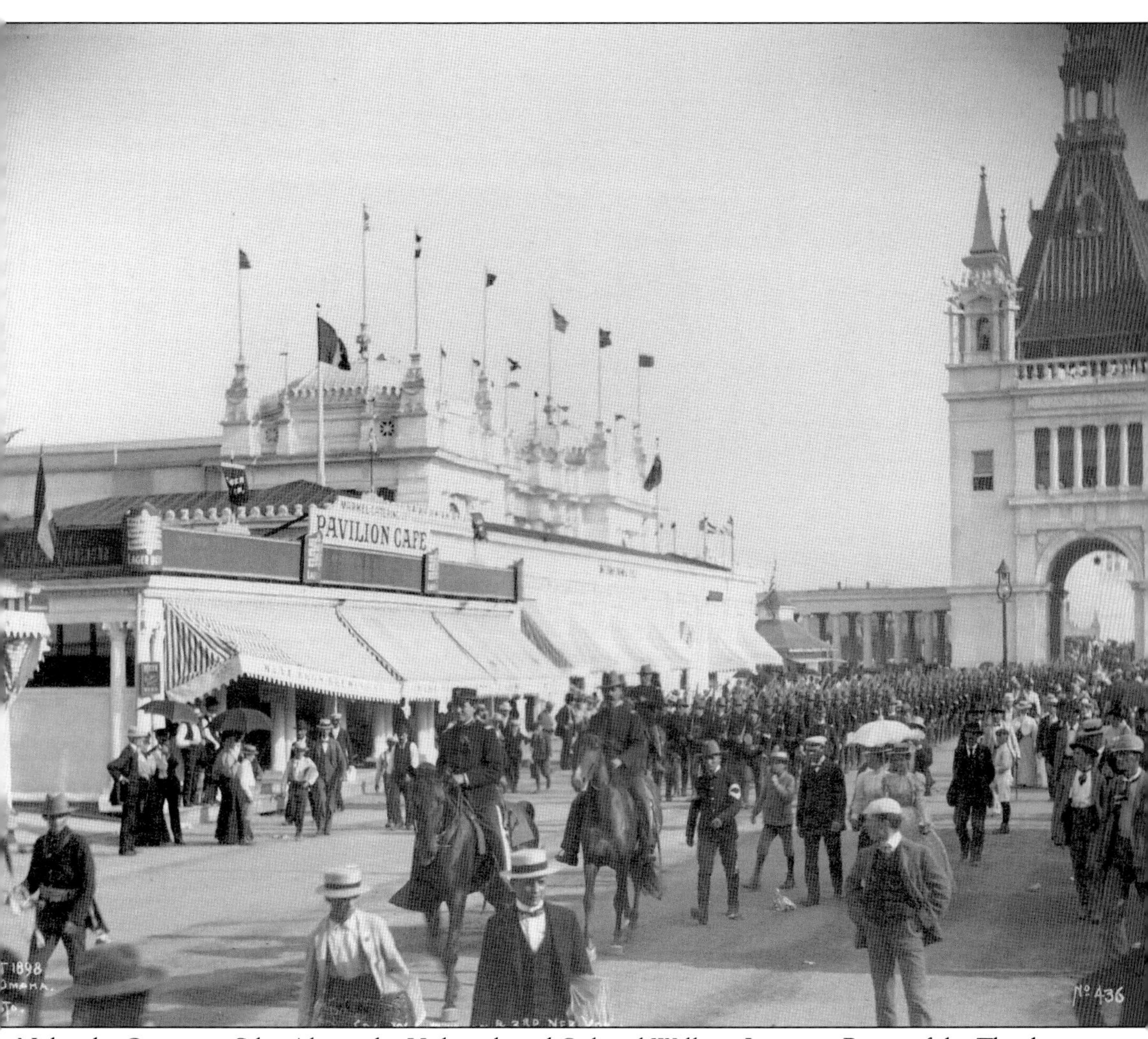

Nebraska Governor Silas Alexander Holcomb and Colonel William Jennings Bryan of the Third Regiment of Nebraska led the parade on Military Day at the exposition. After the parade, the soldiers attended speeches in the auditorium on the exposition grounds, followed by a dinner served by the Ladies' Bureau of Entertainment. (Photo courtesy of Omaha Public Library.)

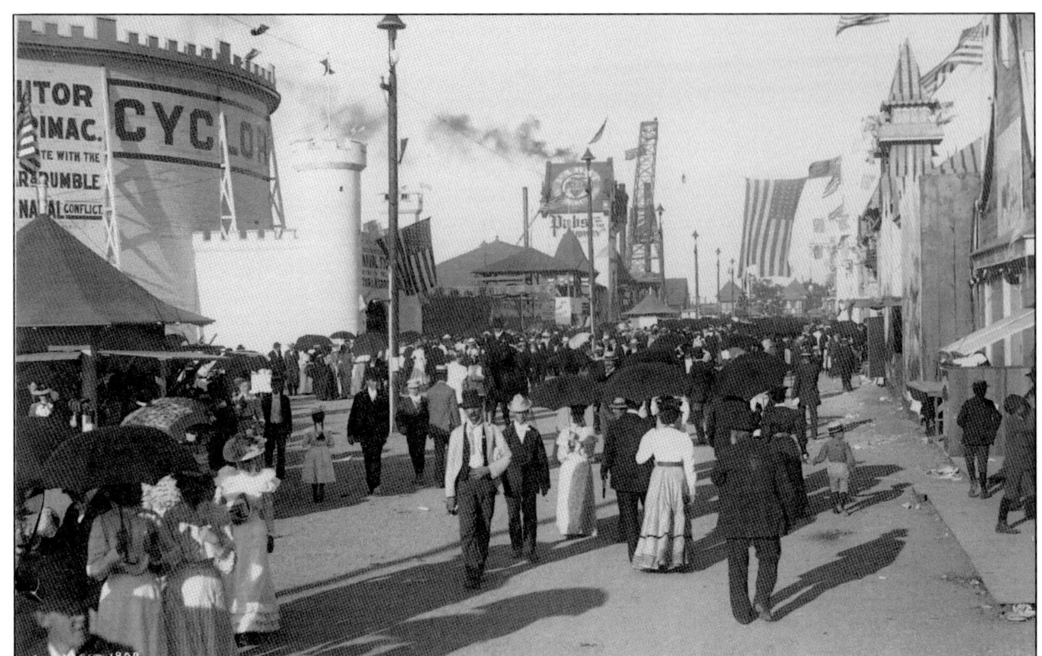

Pictured at the left in this view of the North Midway is the Cyclorama, which, on its interior walls, featured a large-scale depiction of the Civil War battle between the Monitor and the Merrimac. (Photo courtesy of Omaha Public Library.)

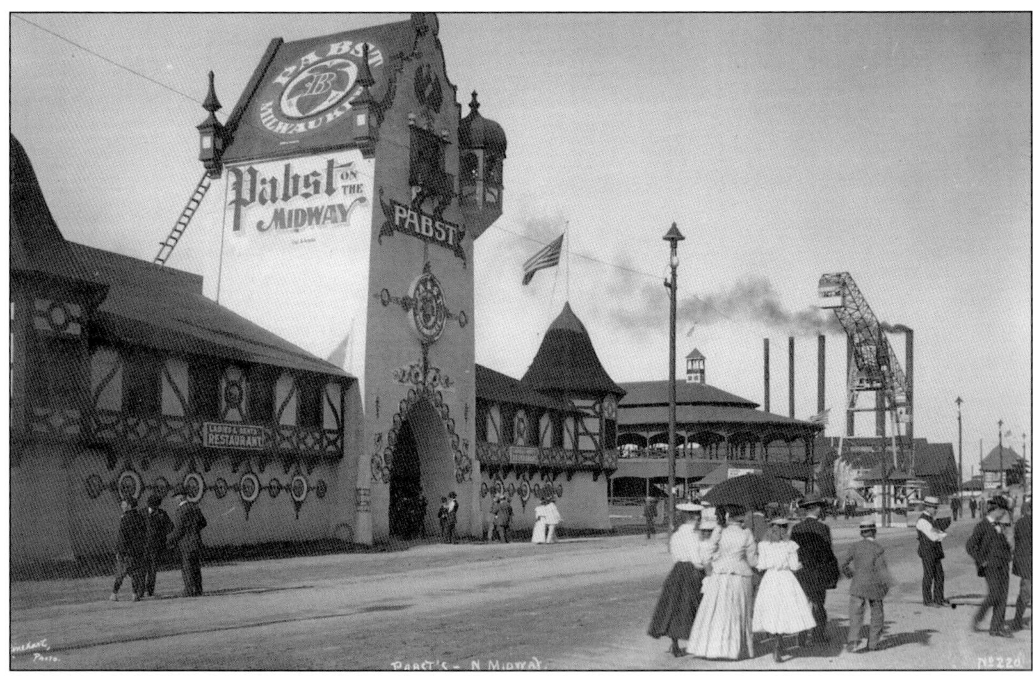

Prominent in this view looking east on the North Midway is the Pabst Pavilion. While the Pabst Pavilion was among the larger restaurants of the exposition grounds, visitors had a variety of choices at mealtime, including light lunches, sandwiches, fruit, and a common array of snacks to treat young and old alike. (Photo courtesy of Omaha Public Library.)

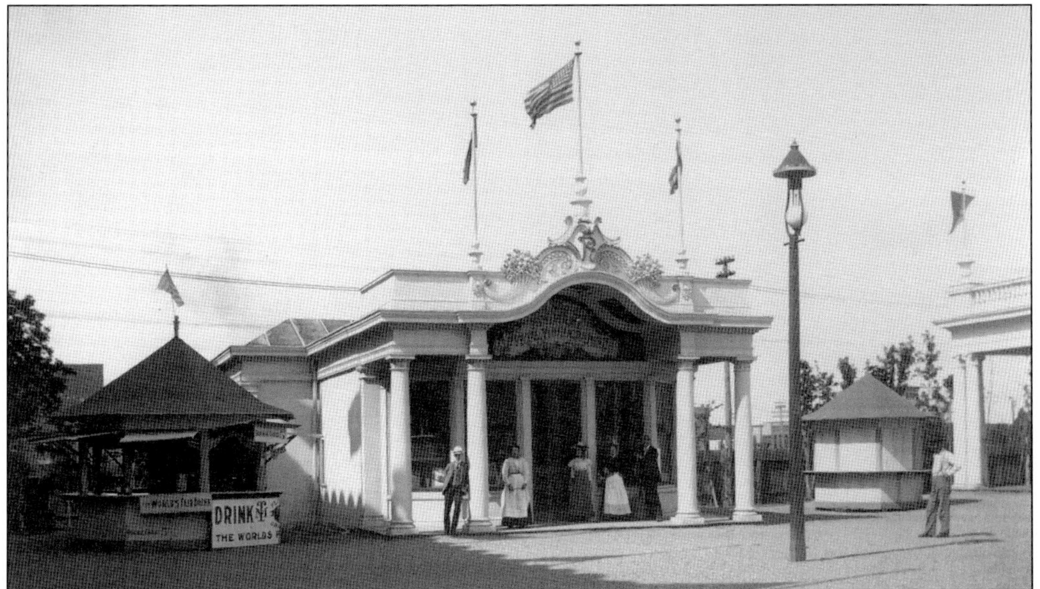

The photographic studio of Frank A. Rinehart, official photographer of the Trans-Mississippi Exposition, was located at the northeast corner of the Grand Court. Exposition visitors, who were charged one dollar to bring their camera on the grounds, were permitted to use the dark room in the facility to load cameras and photographic plate holders. (Photo courtesy of Omaha Public Library.)

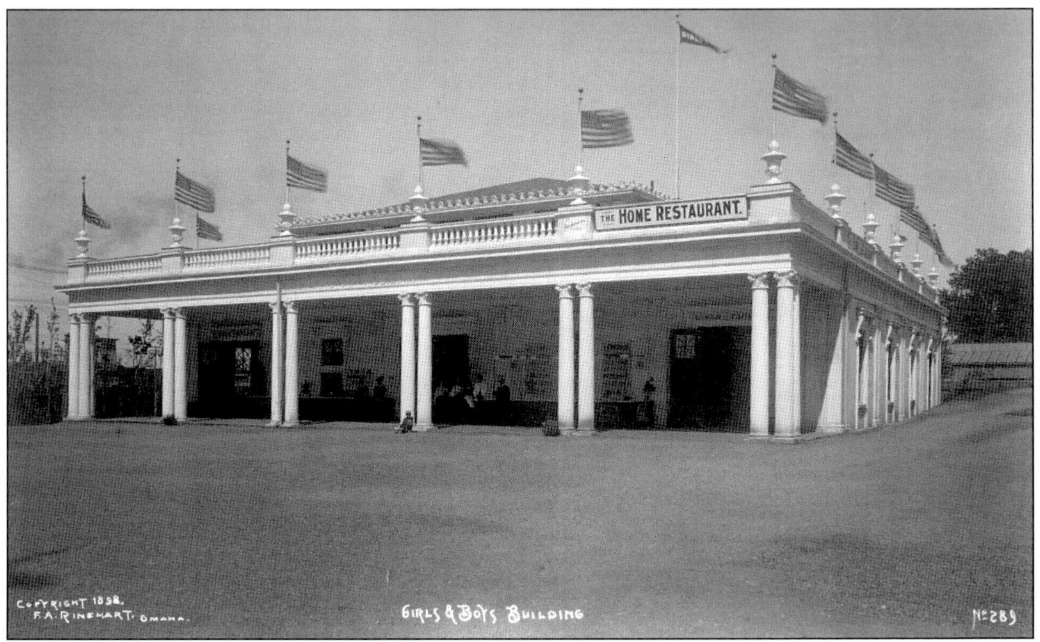

This T-shaped structure, located at the northeast corner of the Grand Court, is the Girls and Boys Building. Along with facilities designed for the care of young children, a modern nursery was on display inside of this building. Parents could also take advantage of the crèche, which cared for approximately 2,000 children during the exposition season. Young children enjoyed a wading pond and recreational sandboxes. (Photo courtesy of Omaha Public Library.)

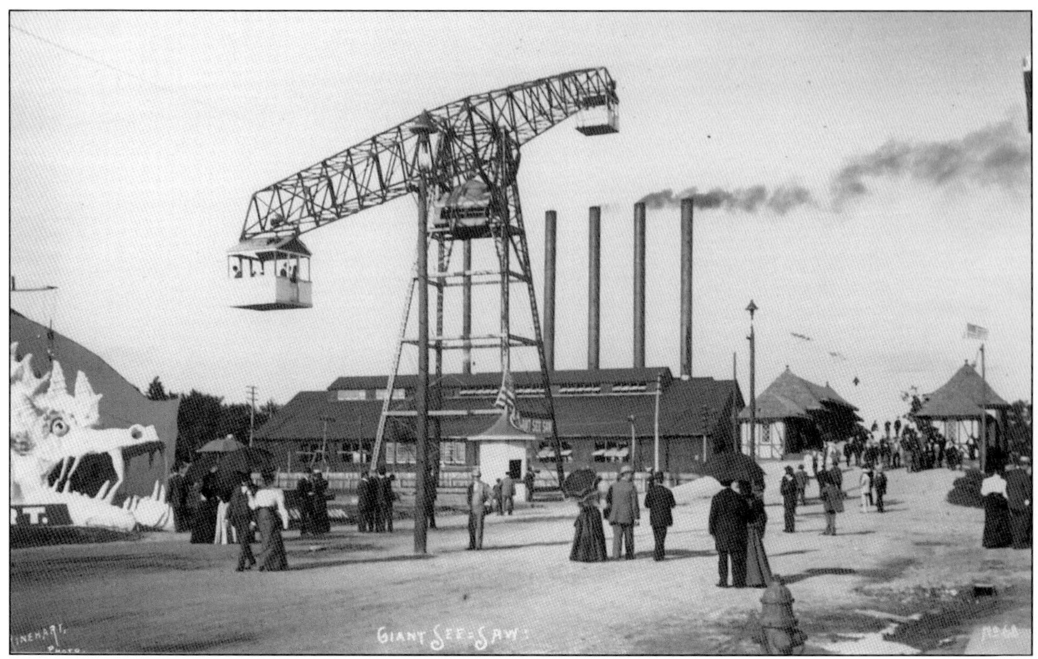

For 10¢, the steel cross beam of the Giant See-Saw elevated riders to a height of nearly 200 feet above the North Midway. Pictured behind the see-saw is the electrical power plant for the exposition and, at the right, the passageway to the East Midway. The unique dragon's head entrance to the "Idols of Art" display is seen at the left of this photograph. (Photo courtesy of Omaha Public Library.)

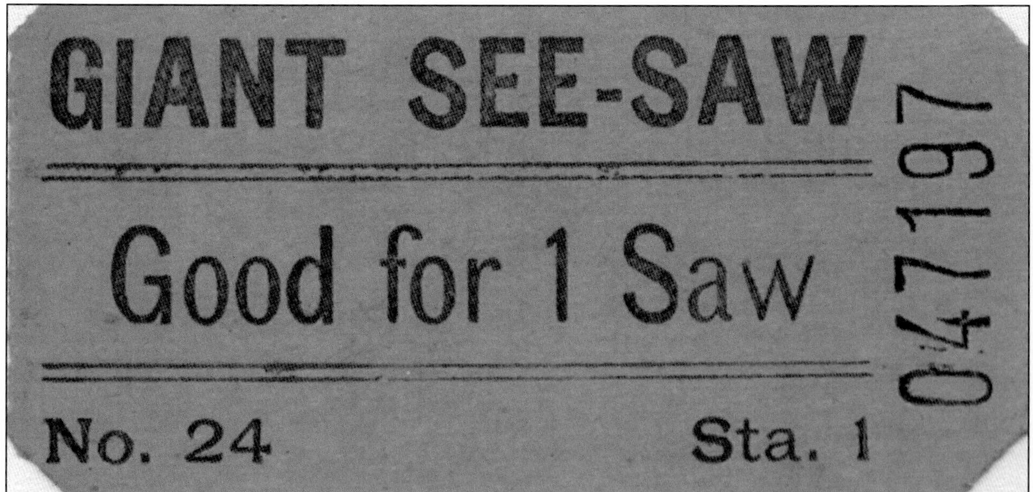

This ticket, entitling a rider to one "saw" on the Giant See-Saw, was of a similar design to many exposition entrance and event tickets. The design and color of exposition tickets sometimes varied within the same day in order to manage fraud and maintain accurate entrance records. (Photo courtesy of Omaha Public Library.)

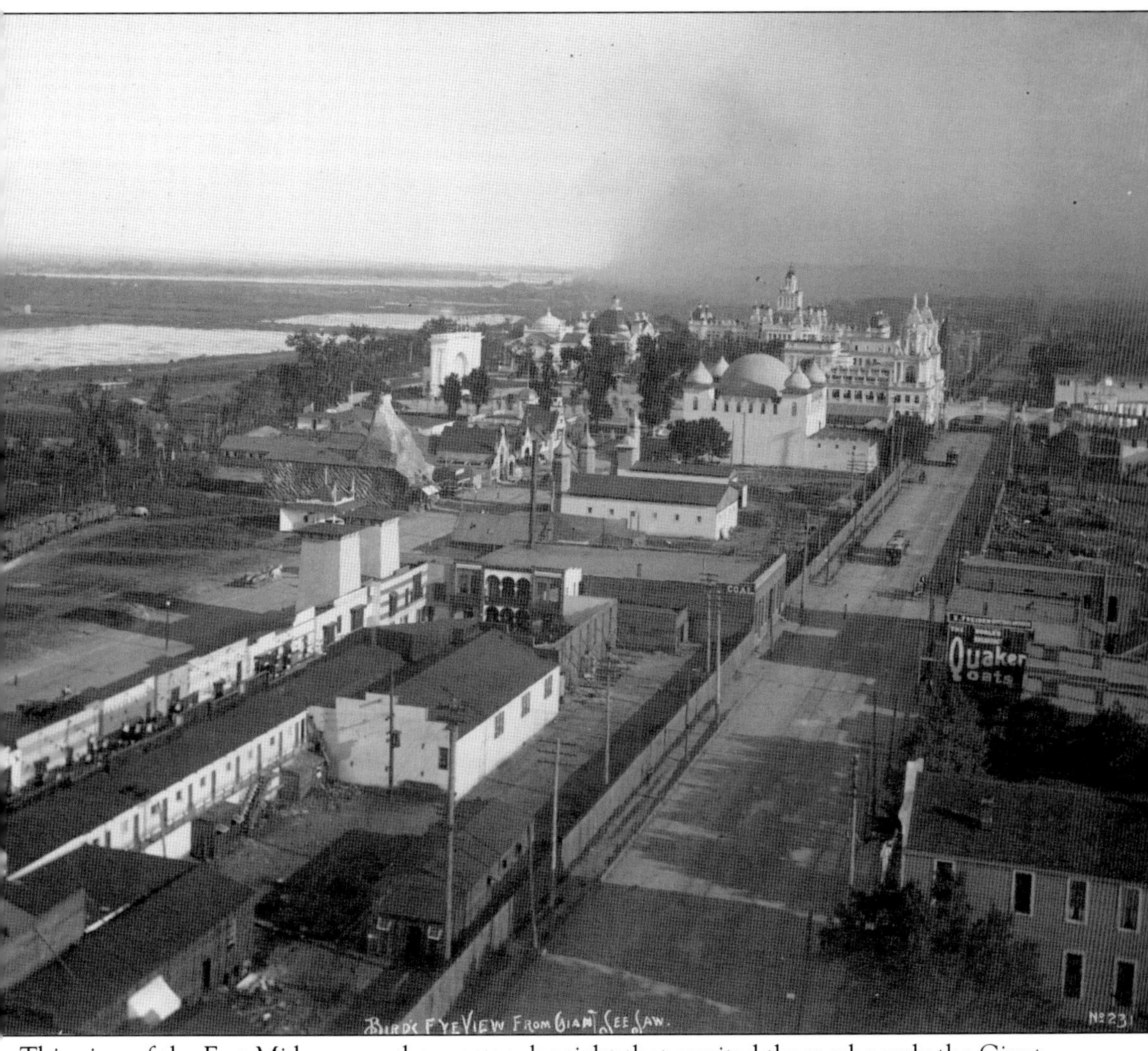

This view of the East Midway was the spectacular sight that awaited those who rode the Giant See-Saw. Prominent visible structures were the Streets of Cairo (lower left), the onion domes of the Moorish Palace and, in the distance, the state buildings of the Bluffs tract. The main roadway pictured at the right is Sherman Avenue. (Photo courtesy of Omaha Public Library.)

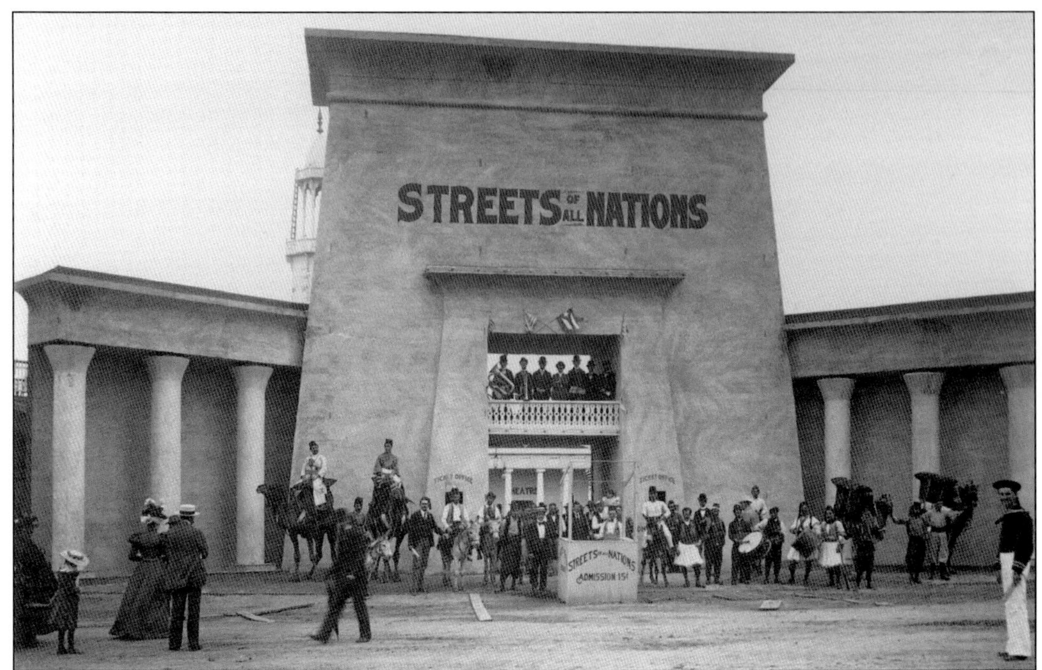

The performers of the Streets of All Nations posed for this picture outside of the attraction. For 15¢, visitors passed through this grand entrance to experience a variety of international-themed displays, events, and concessions. Unique among the sites were international performers in national costumes and animals, such as the camels and donkeys pictured here. (Photo courtesy of Omaha Public Library.)

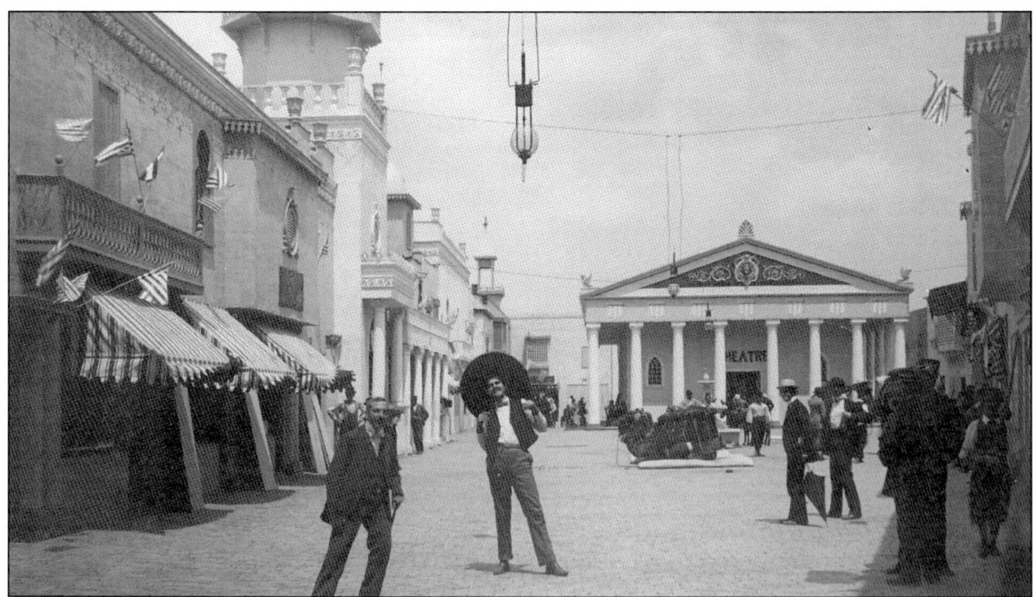

This Turkish minstrel proudly displays his native costume on the Streets of All Nations. Performers, fortune-tellers, and small shops made this attraction of the North Midway a popular diversion for exposition visitors. Resting at the center of the courtyard are camels, which were available to children for rides. (Photo courtesy of Omaha Public Library.)

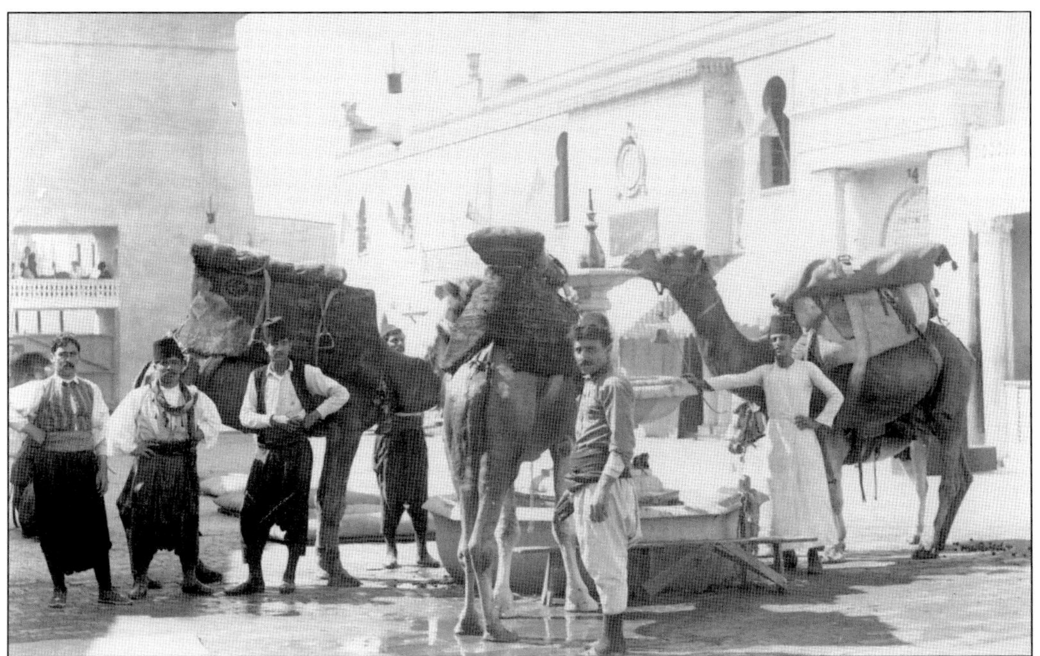

Pictured around the fountain at the center of the courtyard outside of the Streets of All Nations Theatre are camels, saddled for willing riders, and their handlers. The inside of the entrance arch to this area of multiple attractions is visible in the background at the left. (Photo courtesy of Omaha Public Library.)

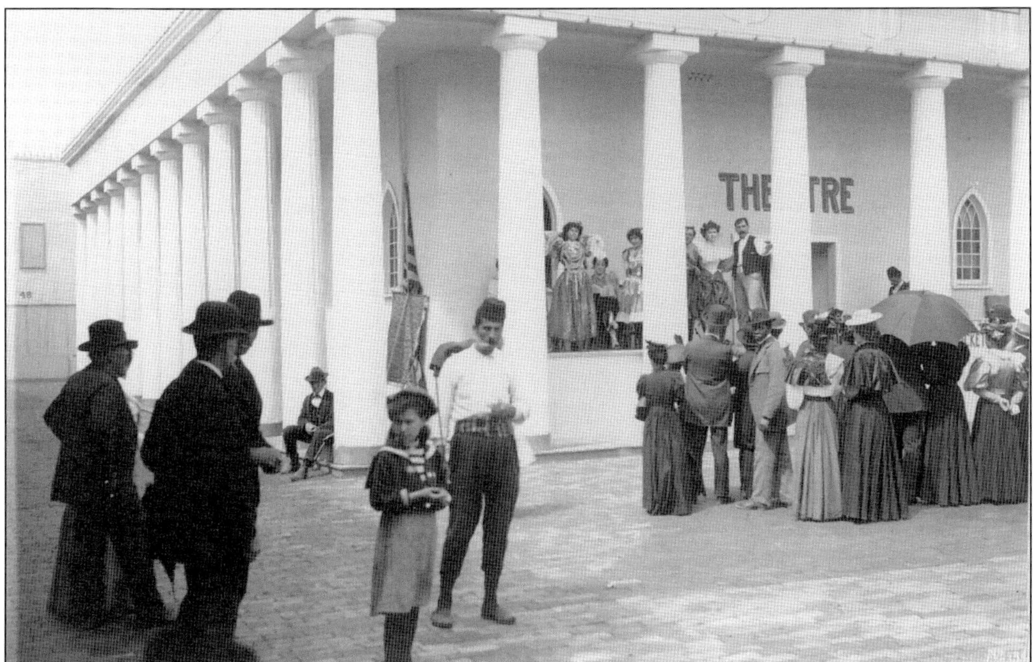

These exposition visitors pass time mingling with performers outside of the Streets of All Nations Theatre. Designed to resemble the Parthenon in Athens, the theater featured vaudeville performers. (Photo courtesy of Omaha Public Library.)

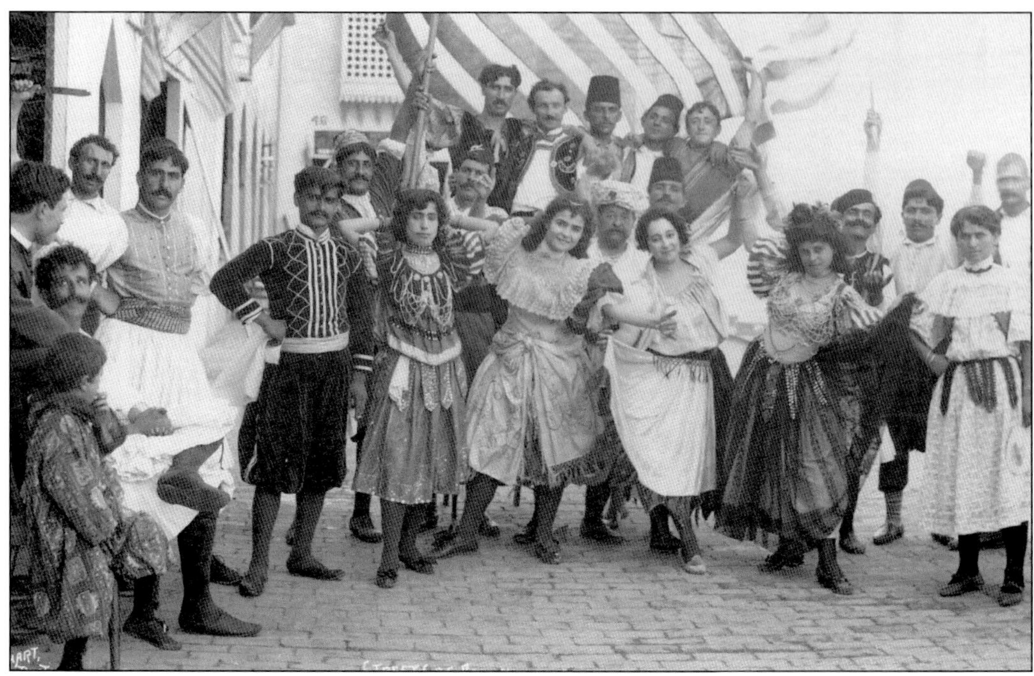

Pictured in the courtyard outside of a restaurant, performers and shopkeepers from the Streets of All Nations appear in native dress and performance costumes. (Photo courtesy of Omaha Public Library.)

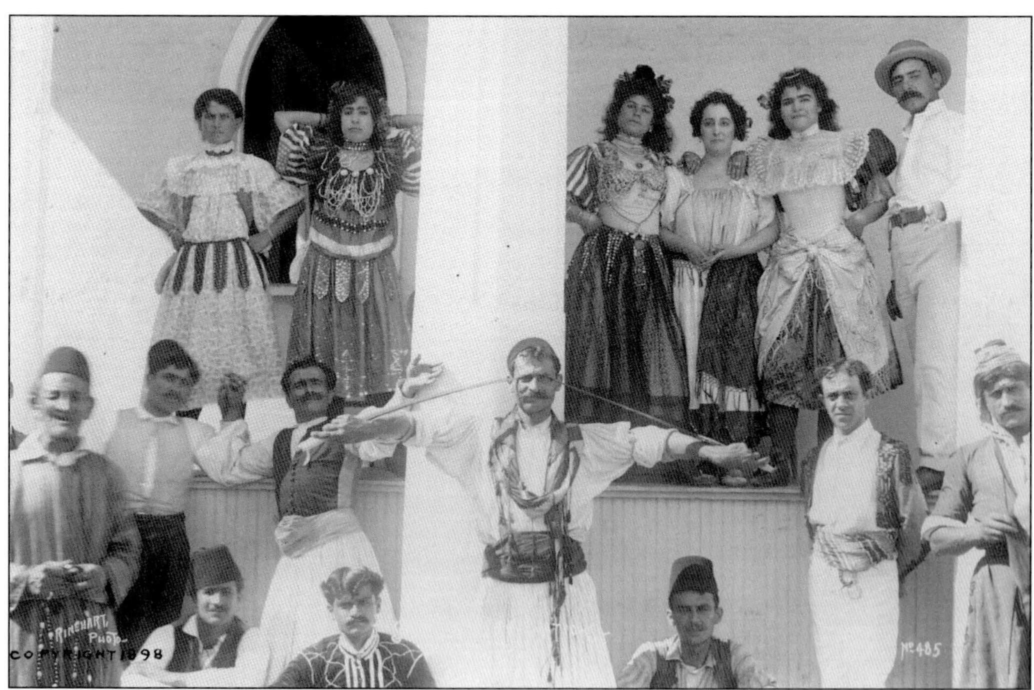

Performers from the Streets of All Nations pose outside of the theater for this photograph. Before each show, a variety of public performances and athletic competitions were presented to those who visited the Streets of All Nations. (Photo courtesy of Omaha Public Library.)

Three years after the exposition, Wilhelm Conrad Röntgen was awarded the Nobel Prize in Physics for his discovery of X-rays. Demonstration of this technology, three years after its discovery, fascinated many exposition visitors. (Photo courtesy of Omaha Public Library.)

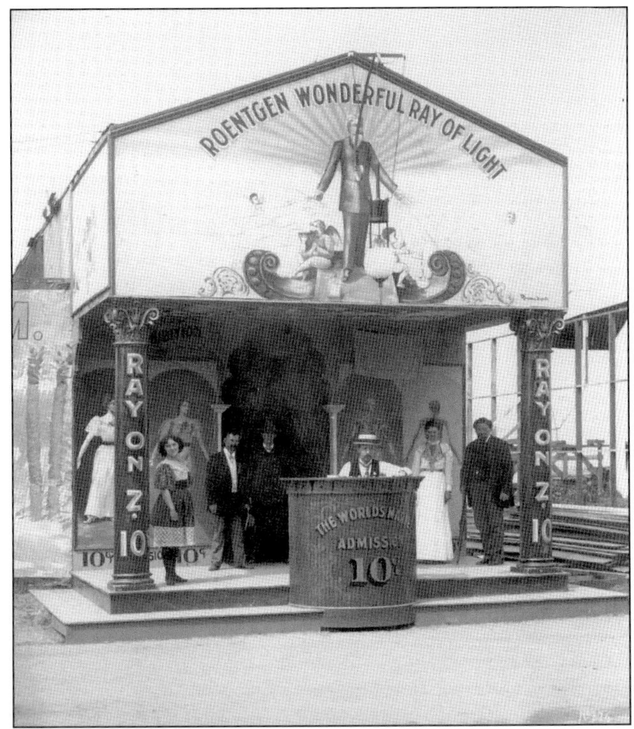

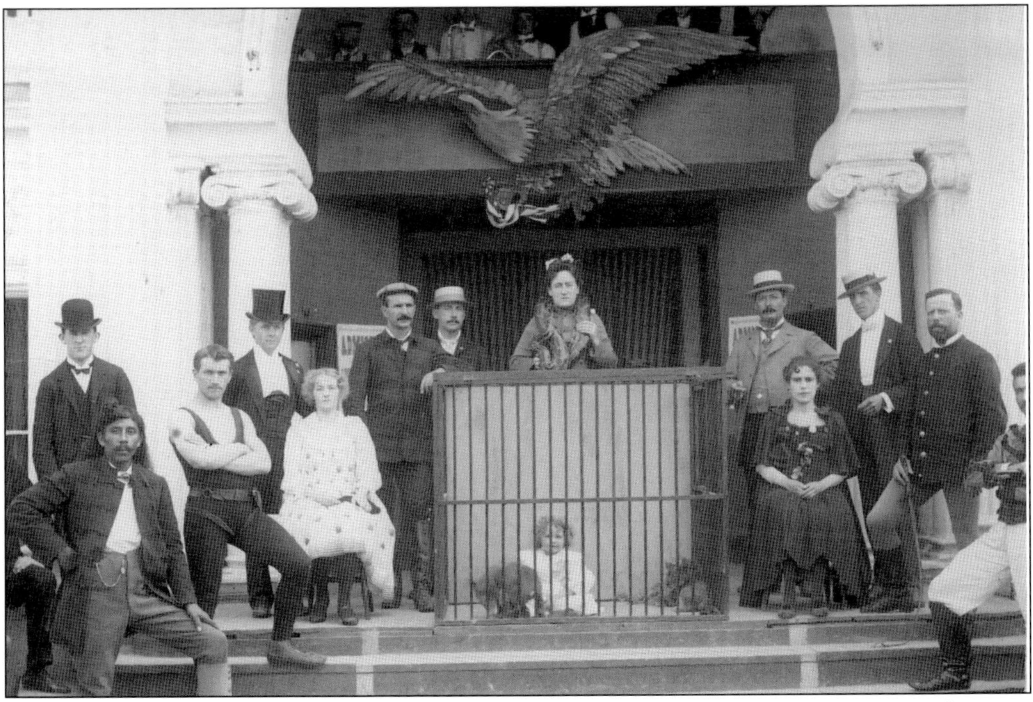

Popular at many expositions of this era, Hagenback's Wild Animal Show was among the most prominent amusements available to midway visitors. Throughout the day, twenty animal trainers gave extensive performances including animals as diverse as elephants, bears, lions, and elks. (Photo courtesy of Omaha Public Library.)

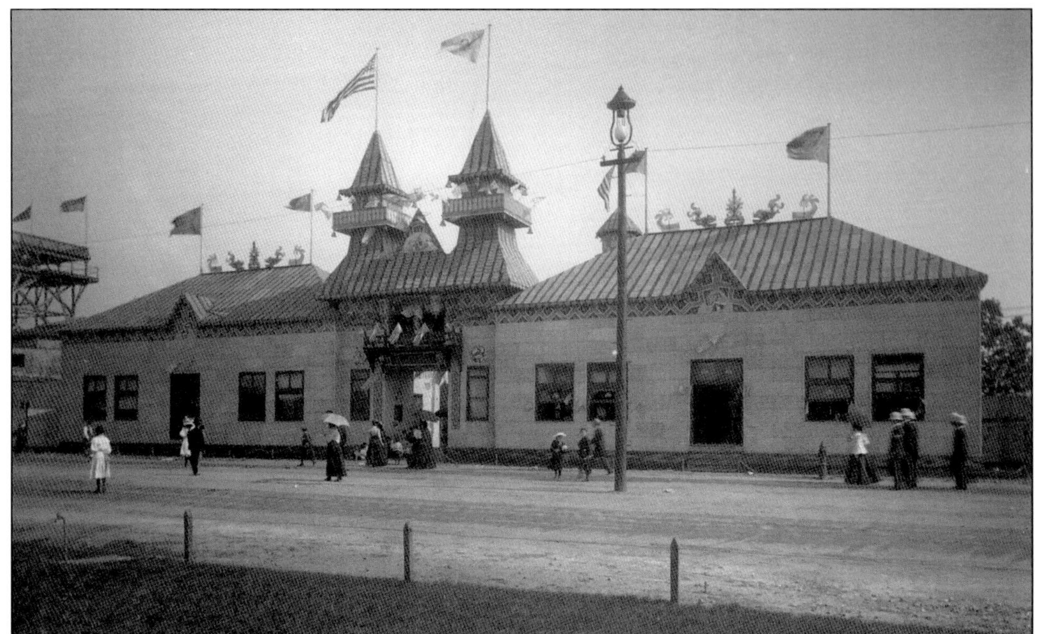

The Chinese Village, a principal attraction of the North Midway, featured performers, crafts and the opportunity to sample the gastronomic delights of this faraway nation. Men and women performed acrobatic feats, while musicians of China added to the international ambience of this exhibit. Available Chinese crafts included works of embroidery, teak wood furniture, and other items popular with exposition visitors. (Photo courtesy of Omaha Public Library.)

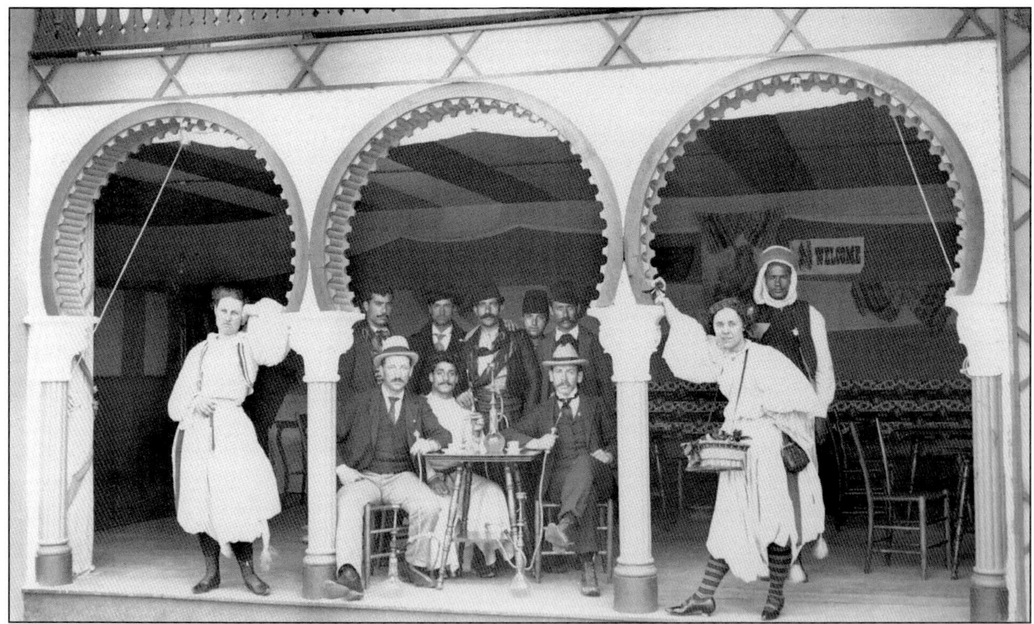

This café, located within the Streets of Cairo, was among the popular relaxation spots at the exposition. Visitors were treated to exotic teas and treats, as well as a visual display of native design, including the horseshoe-shaped arches common in Islamic architecture. (Photo courtesy of Omaha Public Library.)

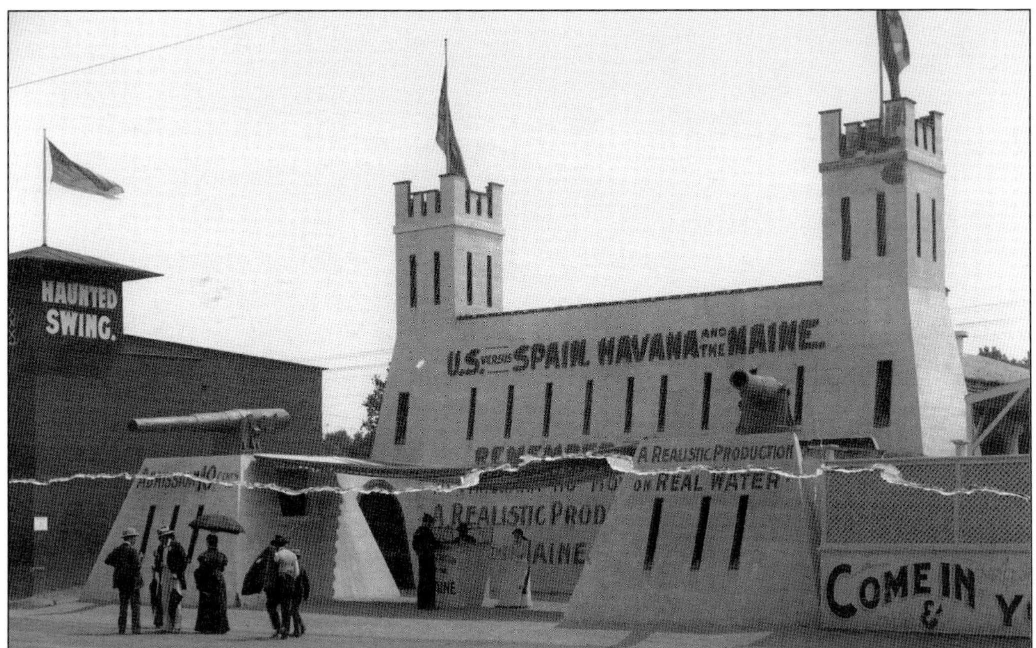

On February 15, 1898, the *U.S.S. Maine* exploded in Havana Harbor, resulting in a call for war against Spain. This historical production of the North Midway encouraged visitors to "Remember the Maine!" Though significant preparations for the exposition had been set in motion at the time of this incident, exposition organizers took pause when considering how this event would affect the success of their undertaking. (Photo courtesy of Omaha Public Library.)

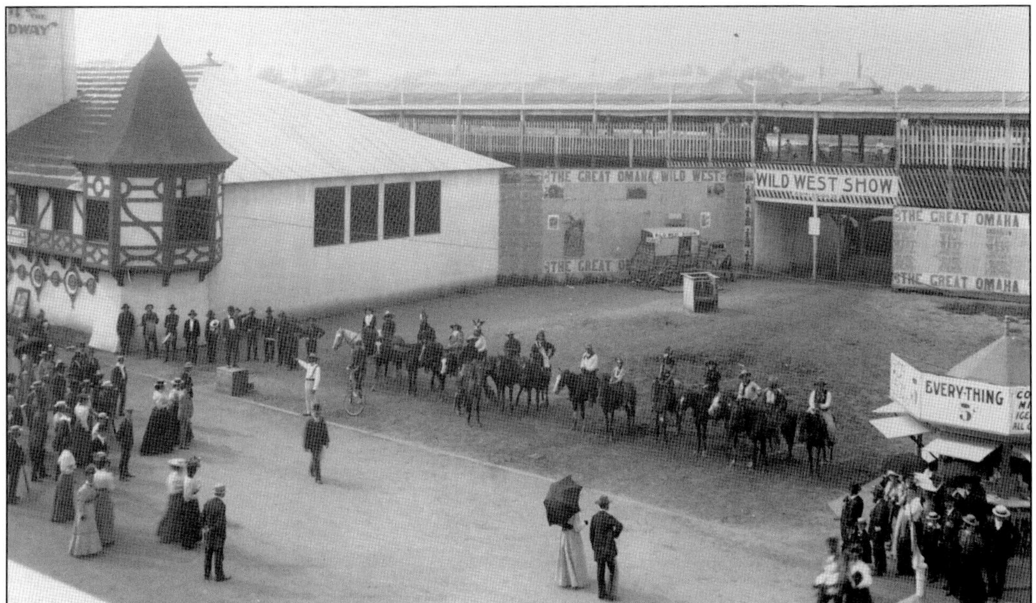

William F. "Buffalo Bill" Cody operated the Wild West Show that appeared at the Trans-Mississippi Exposition. For 25¢, visitors saw demonstrations of rodeo skills and a mock attack on a stagecoach. Cody led the parade at his show on "Cody Day" at the exposition, which took place on August 31, 1898. (Photo courtesy of Omaha Public Library.)

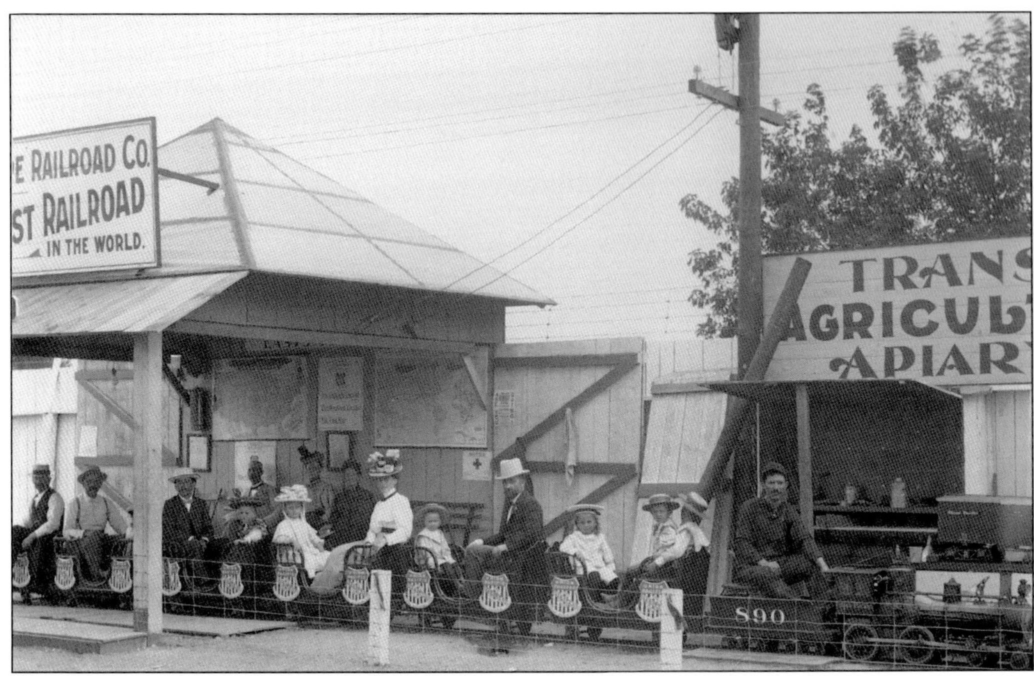

The "smallest railroad in the world" was popular with young and old alike at the Trans-Mississippi Exposition. Ten passenger cars, each containing seats for two riders, featured the familiar shield of the Union Pacific emblem. Posted on the walls of this "station" were maps and advertisements for the Union Pacific Overland Route. (Photo courtesy of Omaha Public Library.)

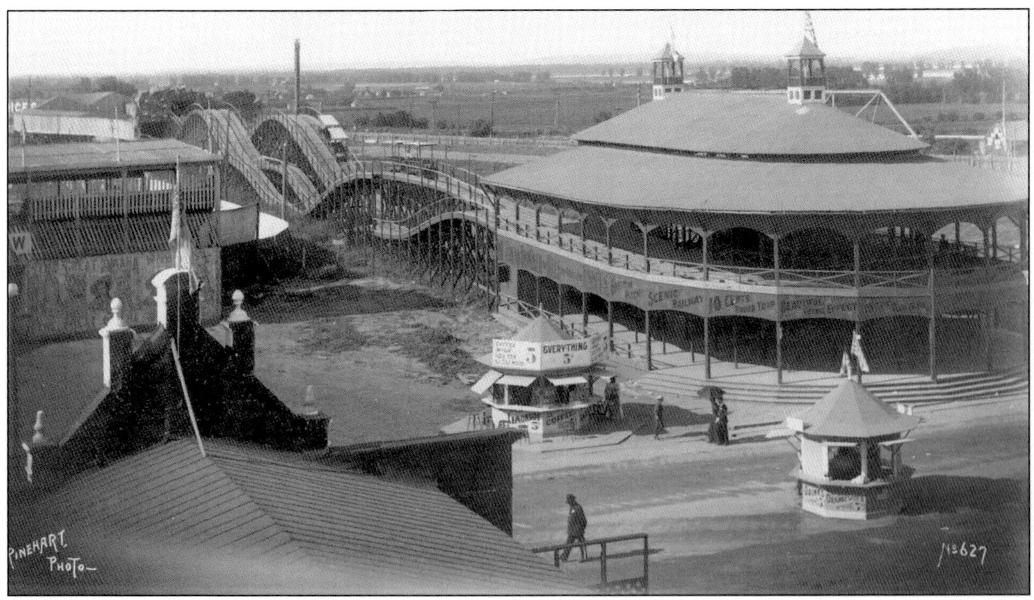

For 10¢, the Scenic Railway of the North Midway promised a "round trip over the hills and through the tunnels" along with a "beautiful effect in the tunnel." The three-minute ride, journeying the distance of the 4,000-foot track, featured two tunnels, one depicting the American military conquest of Manila, and a second through a suspenseful area in complete darkness. (Photo courtesy of Omaha Public Library.)

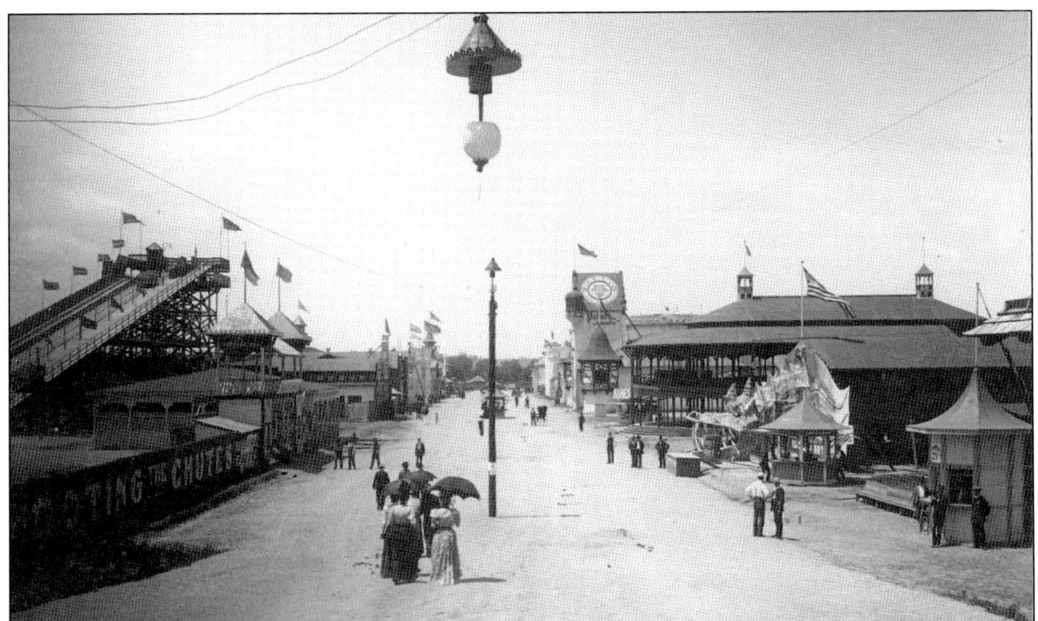

Forty-five star United States flags fly atop the peaks of most structures in this view of the North Midway. Pictured on the left is the incline of the popular "Shooting the Chutes" ride. On the right is the two-story, covered start of the scenic railway. Beyond the railway is the Pabst Pavilion, offering exposition visitors meals and the opportunity to sample the familiar brand of beer. (Photo courtesy of Omaha Public Library.)

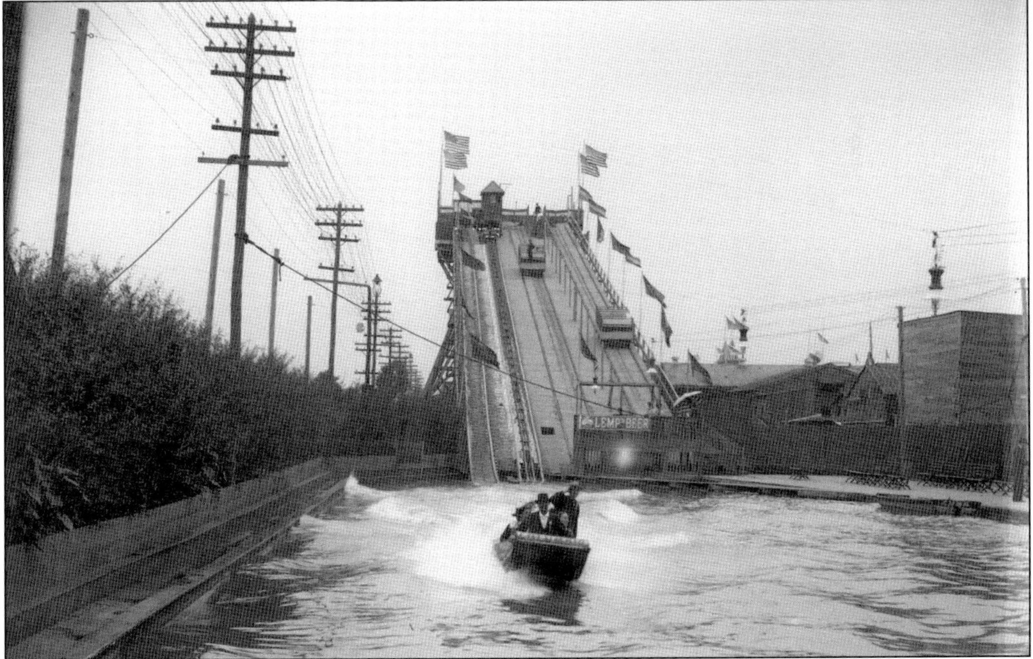

From a height of 108 feet, eight passengers descended 300 feet from the platform at the top of the Shooting the Chutes ride to the lake below. The speed of the descent caused the craft to jump as it propelled riders across the lake. (Photo courtesy of Omaha Public Library.)

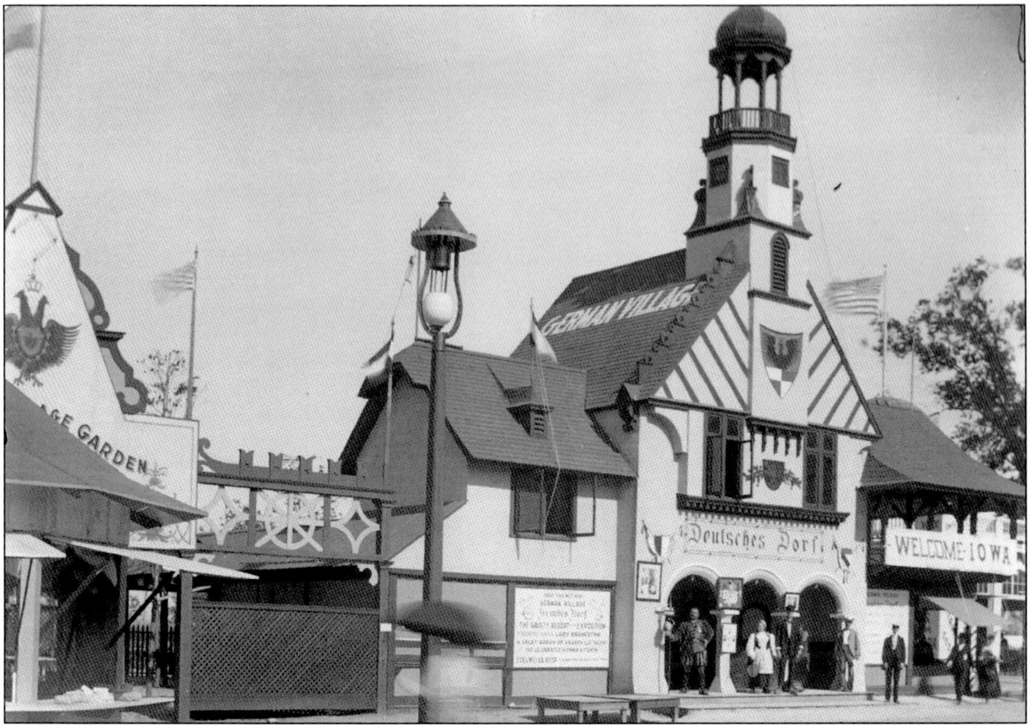

The Deutsches Dorf, or German Village, reproduced the Bratwurst Cloecklein, an historic inn of Nuremberg. Exposition "special days" recognizing specific ethnic groups, such as German Day, held on October 18, 1898, were of particular importance for proprietors of this type of attraction. (Photo courtesy of Omaha Public Library.)

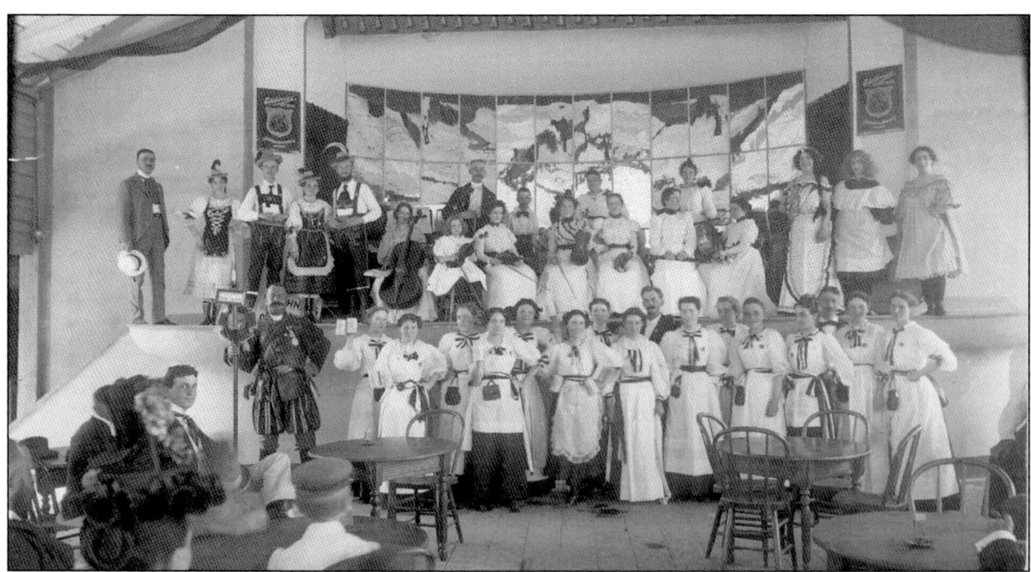

"Kirchner's Famous Lady Orchestra" was the featured attraction among a variety of vaudevillian performers in the German Village. Along with the entertainment, patrons could enjoy the "Vienna Kitchen" and, if they desired, a stein of the house favorite, Edelweiss. (Photo courtesy of Omaha Public Library.)

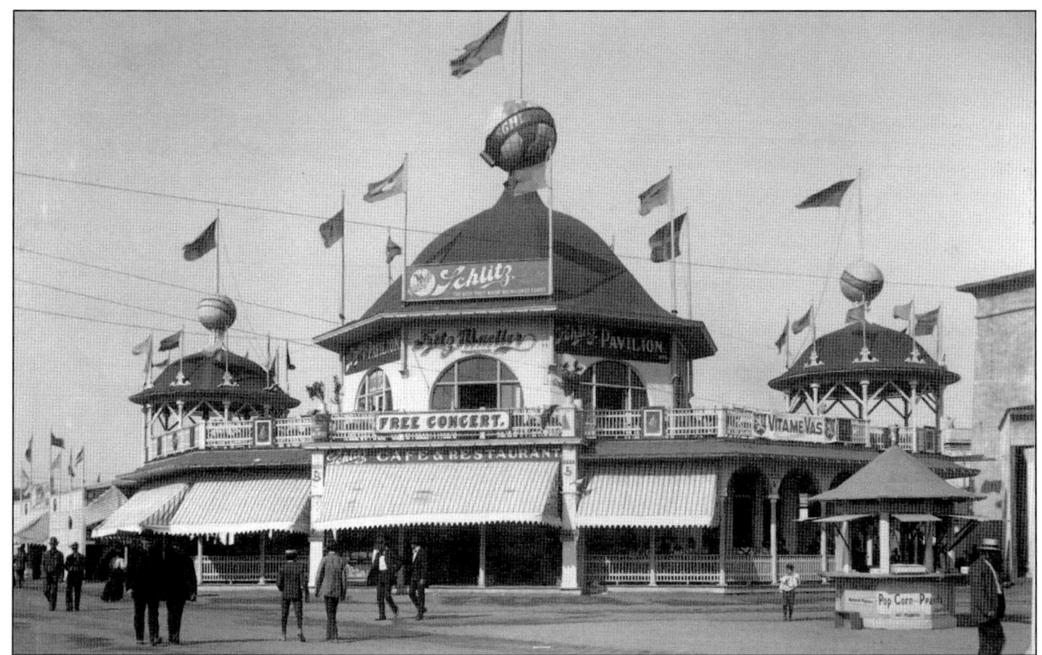

Fritz Mueller served "the beer that made Milwaukee famous" in the Schlitz Pavilion on the North Midway. Adding to the atmosphere were Swiss barmaids and a Hungarian band. (Photo courtesy of Omaha Public Library.)

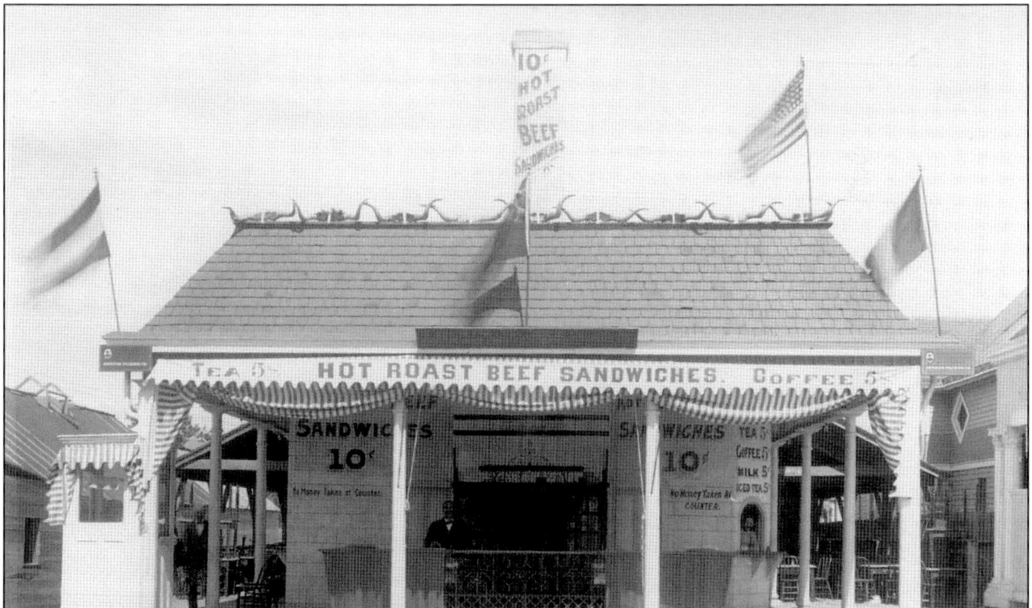

Hot roast beef sandwiches were available for 10¢ from this stand to those who strolled the exposition midway near the Grand Plaza. For an additional 5¢, one could add tea, coffee, milk, or iced tea. (Photo courtesy of Omaha Public Library.)

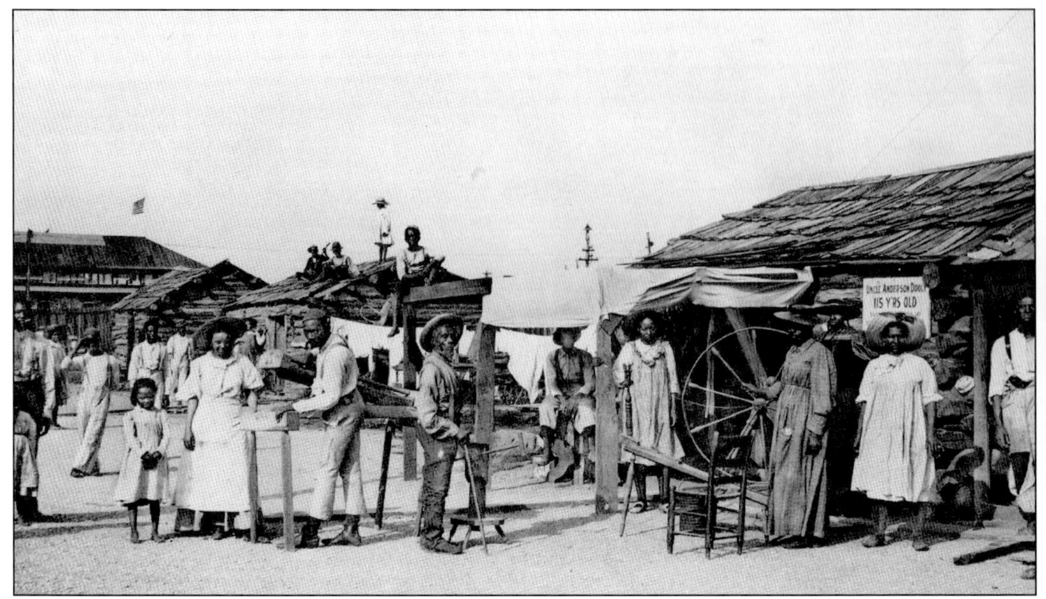

Exhibiting life on an antebellum cotton plantation, the Old Plantation featured 12 log cabins, working cotton gins, and African-Americans engaged in processing the principle agricultural product of the South. A theater within the Old Plantation exhibit presented short plays, dance, and music, which were regarded as unique by those who had never visited the land of the former Confederacy.

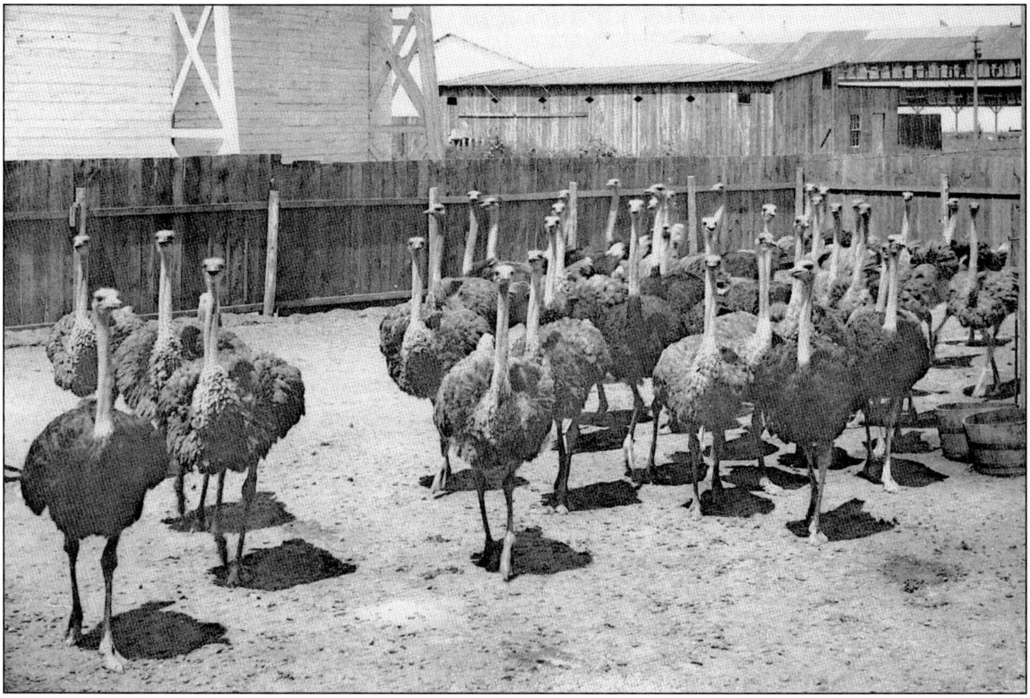

The ostrich farm of the West Midway provided a unique display of these animals from California. Approximately 100 ostrich were presented in this exhibit, familiar to many expositions of the era.

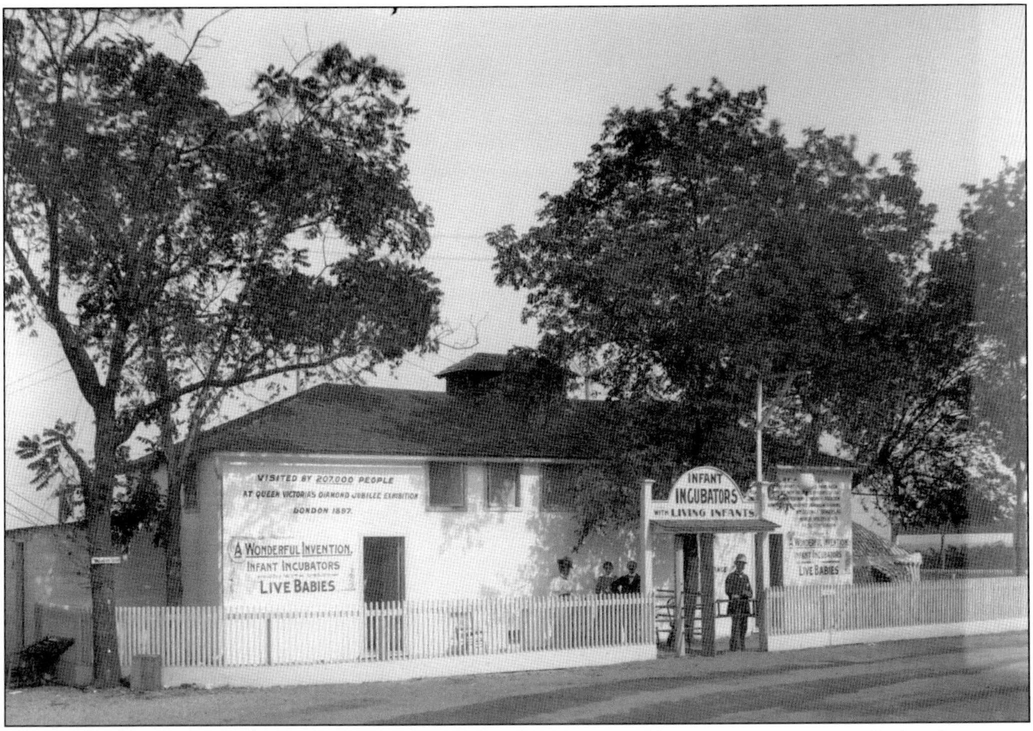

The Infant Incubators Building at the Trans-Mississippi Exposition presented the first-ever display of this technology within the United States. As a developing technology, sponsors promoted their exhibit with a message on the outside of the building boasting that an 1897 display in London had been "visited by 207,000 people at Queen Victoria's Diamond Jubilee." (Photo courtesy of Omaha Public Library.)

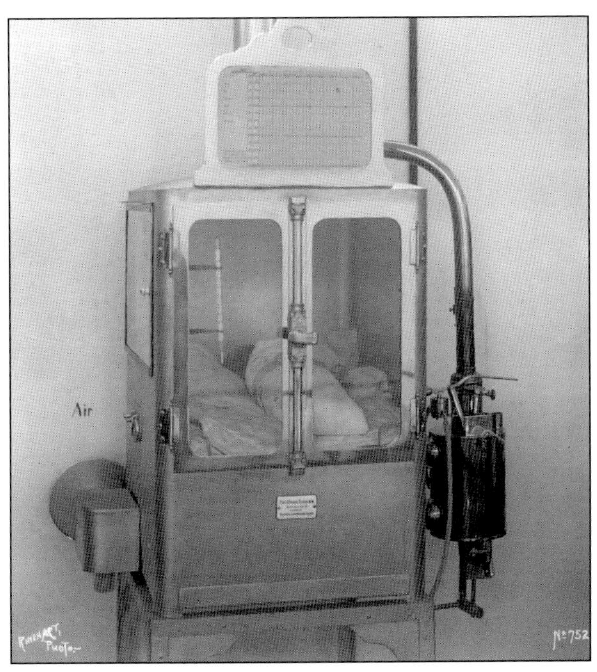

German doctor Martin Couney displayed infant incubators providing support to premature babies. Fresh air from outside was warmed and moistened before reaching a live infant in the incubator chamber. Dr. Couney continued to sponsor incubator exhibits at other fairs and expositions, until the technology became commonplace in hospitals. (Photo courtesy of Omaha Public Library.)

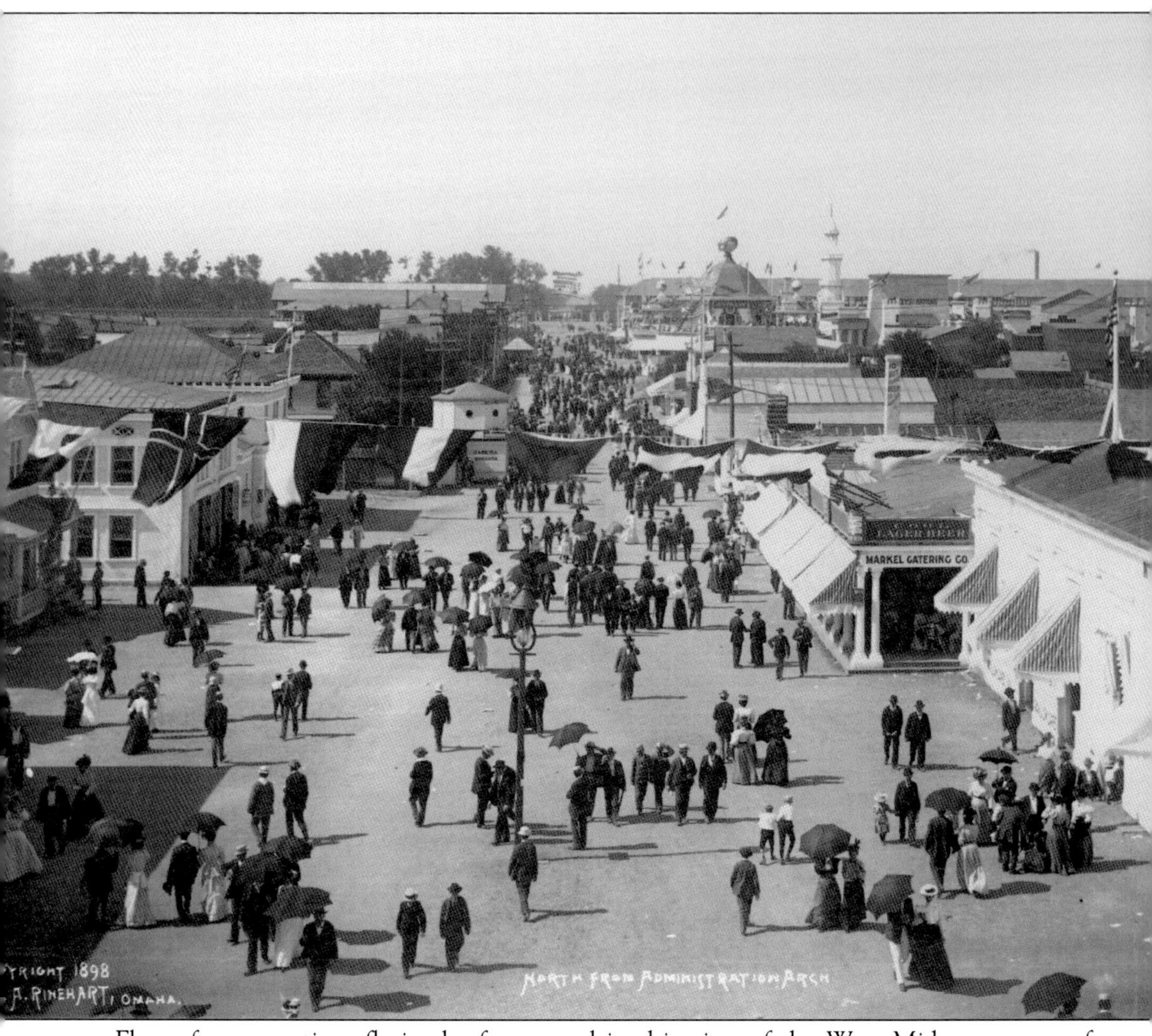

Flags of many nations fly in the foreground in this view of the West Midway, as seen from the Administration Arch. The large, white International Building pictured at the right offered 18,583 square feet of exhibition space. Among the nations featured in this building were Austria, Canada, Denmark, France, Germany, Great Britain, Italy, Mexico, Russia, and Switzerland. An exhibit of Hawaiian farm products and traditional island culture was also on display in this building. (Photo courtesy of Omaha Public Library.)

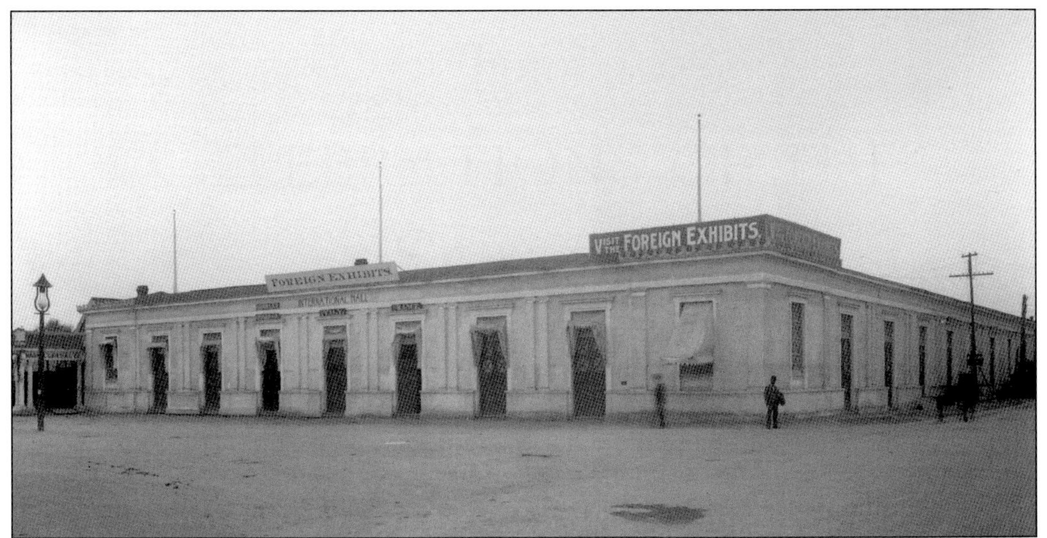

The International Building, located on the West Midway, immediately north of the Administration Arch, was a popular destination for exposition visitors wishing to purchase a unique item of foreign manufacture. Cigars and jewelry from Mexico, Limoges porcelain and fine furniture from France, rugs from Sweden, and a variety of other items from Europe and the Far East were available to interested shoppers. (Photo courtesy of Omaha Public Library.)

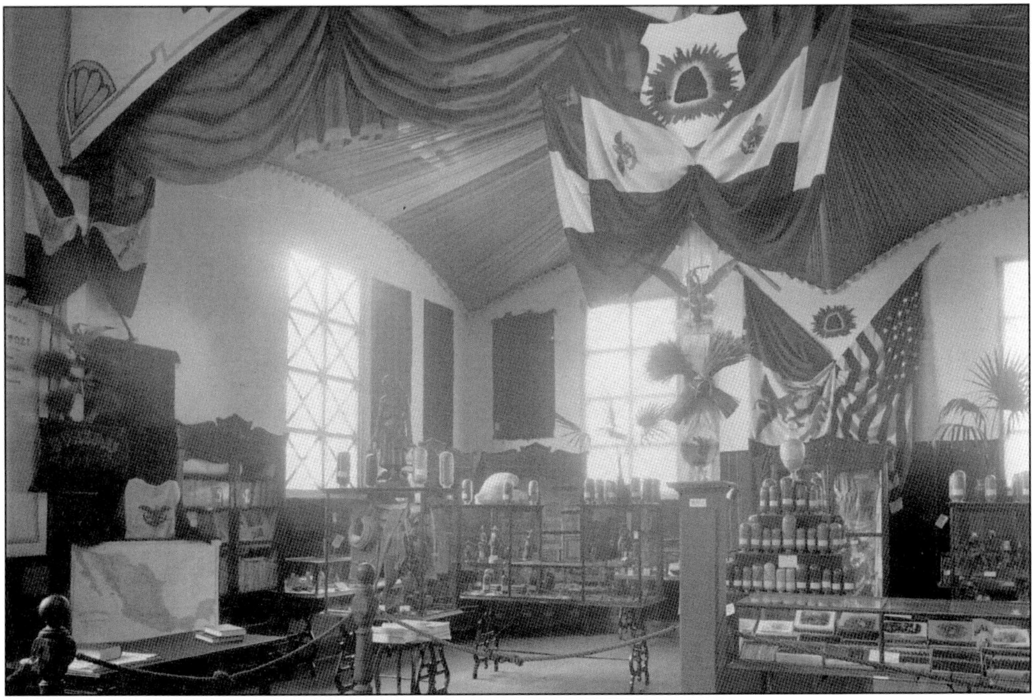

Featured in this display from Mexico were agricultural products, including coffee, corn, lemons, oranges, tobacco, and wheat. Mexican manufactured goods on display included fine linen, silk, and lace, as well as a collection of small statues. A fine historic exhibit of Aztec items made from bone, copper, and stone was also presented in this display from the International Building of the exposition. (Photo courtesy of Omaha Public Library.)

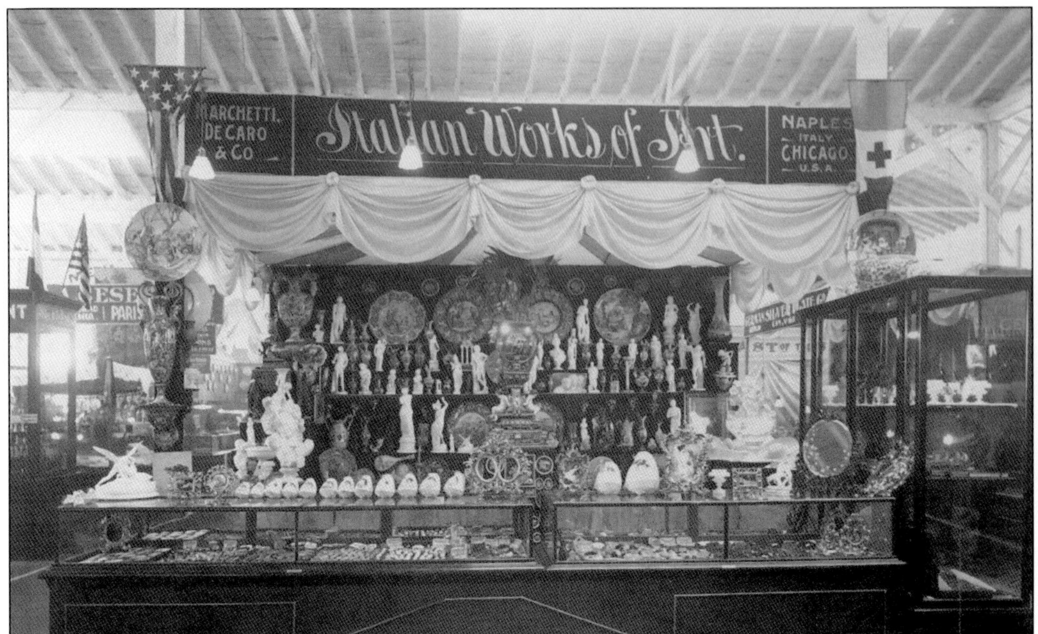

The products of Marchetti, DeCaro and Company from Naples, Italy, were featured in this display from the International Building. Included among the art products of this company were jewelry, frames, plates, figurines, and vases. Many other treasures of Italian art, including some dating back to ancient Rome, were presented under the auspices of the Italian government. (Photo courtesy of Omaha Public Library.)

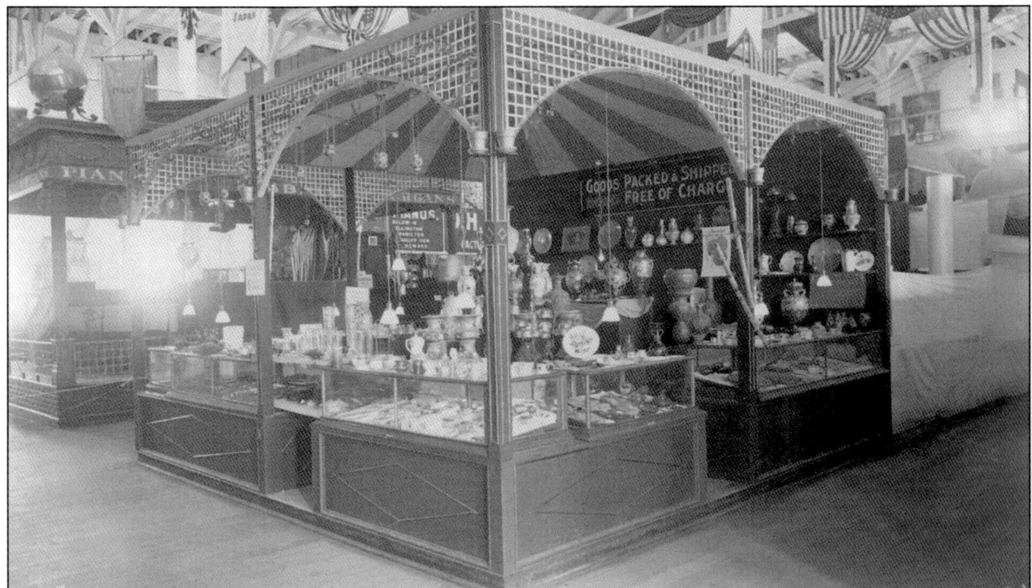

The International Building included this booth featuring the glassware and china products of George D. Sosnowski and Company. An importer of French goods, Sosnowski displayed glasses, vases, plates, and other pieces of fine china. The International Building housed many displays of artistic items, as well as natural products and manufactured goods. (Photo courtesy of Omaha Public Library.)

Four

Celebrations of the Season

Designated to honor specific states, communities, societies, and industries, the special days of the Trans-Mississippi Exposition season attracted many local, as well as out-of-town, visitors, resulting in record attendance figures. Culminating with the celebration of Peace Jubilee Week, including the visit of President William McKinley, and Omaha Day on the final day of the exposition, special days often exceeded the expectations of organizers.

Including days such as Swedish American Day, June 24; Kansas City Day, August 6; Shriners Day, September 14; and Livestock Day, October 3, special celebrations were a key component of a plan designed to attract and welcome visitors to the exposition. A successful effort to bring conventions to Omaha during the exposition brought many additional visitors to the celebration.

Special days featured the speeches of Trans-Mississippi officials, leading politicians, industry captains and religious leaders. Programs also commonly included instrumental and vocal music and events appropriate to the nature of the celebration.

On Iowa Day, September 21, 1898, speaker Robert C. Cousins remembered the Trans-Mississippi celebrants, stating that, "Into the opening gateway of the twentieth century, hand in hand, shall spring our king of commerce and the queen of industry, the Sphinx-eyed scientist and his bride of art, the sturdy son of agriculture and the dreaming child of song, and their thought and toil and song shall honor and inspire the human race and make our country great—essentially, exquisitely, magnificently great."

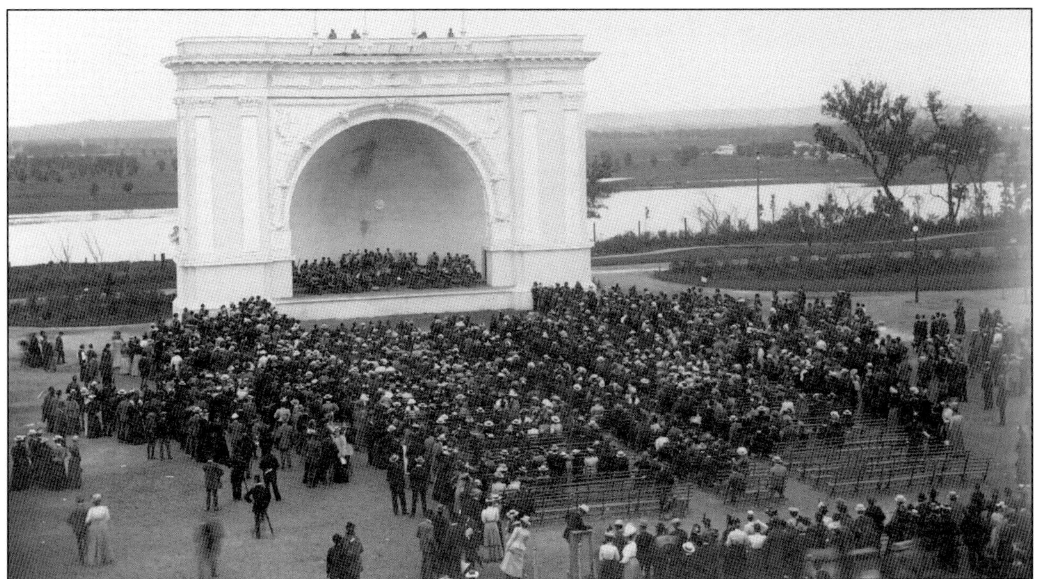

The Music Pavilion, located at the Grand Plaza, east of the Grand Court, was the scene of much celebration during the exposition. Among the many musical presentations and historic events that took place here was the address of President William McKinley on President's Day, held on October 12, 1898. (Photo courtesy of Omaha Public Library.)

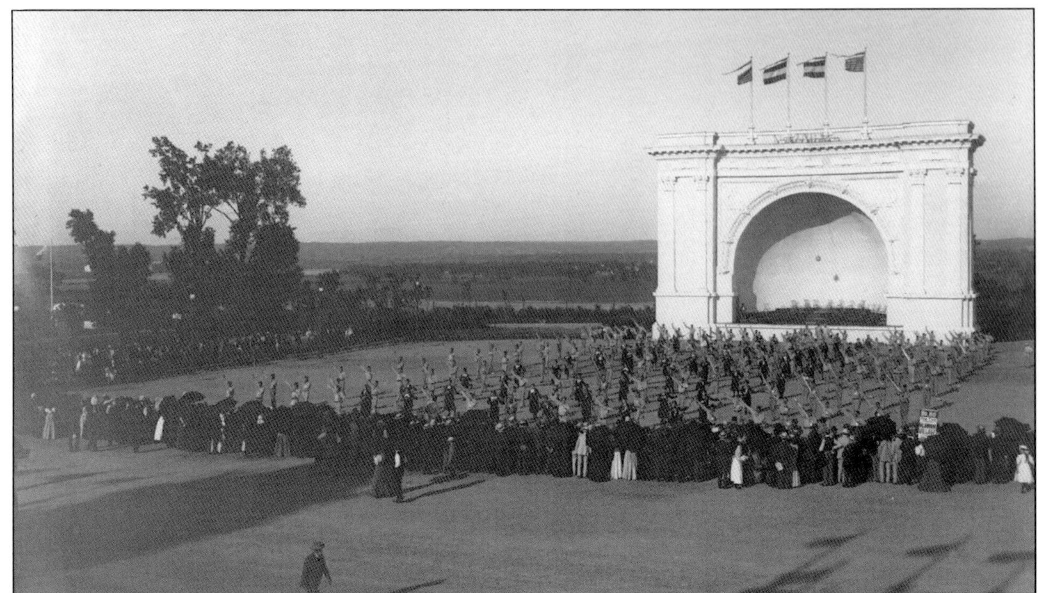

On June 30, 1898, at 6:00 p.m., the Omaha Turners Society sponsored this grand drill featuring approximately 400 participants performing their skills on the Grand Plaza. (Photo courtesy of Omaha Public Library.)

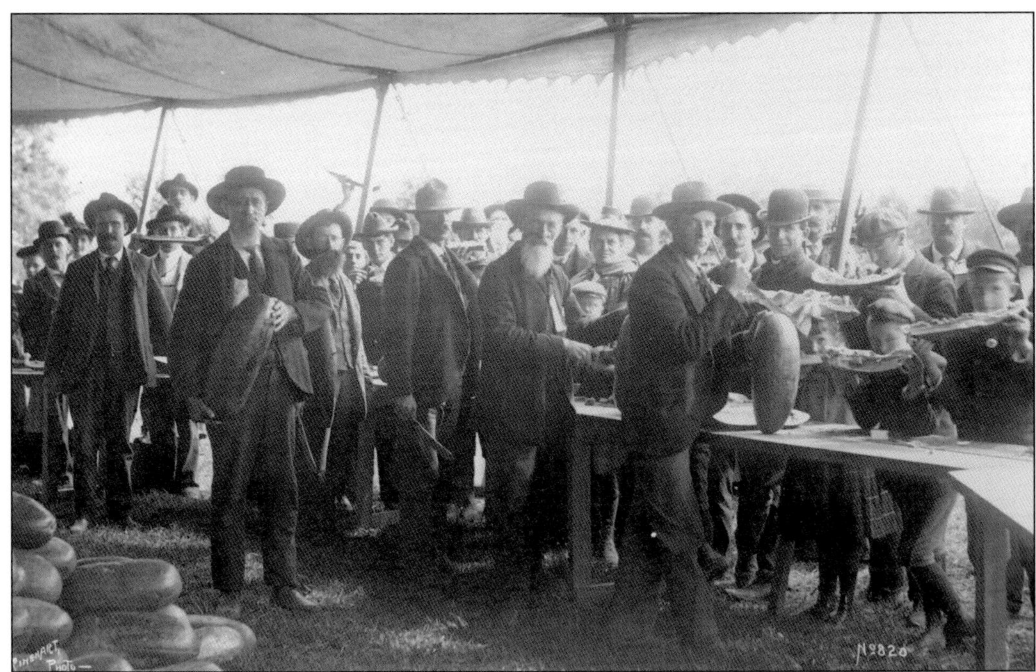

The free distribution of watermelon slices was an occasional, and very popular, practice at the exposition. Watermelons from Rocky Ford, Colorado, the self-proclaimed "Sweet Melon Capital of The World," competed with events such as Texas Melon Day on July 1, 1898—during which an estimated 10,000 visitors were served, including the two young boys pictured at the right, who eagerly accepted a quarter-slice from one of the free melons. (Photo courtesy of Omaha Public Library.)

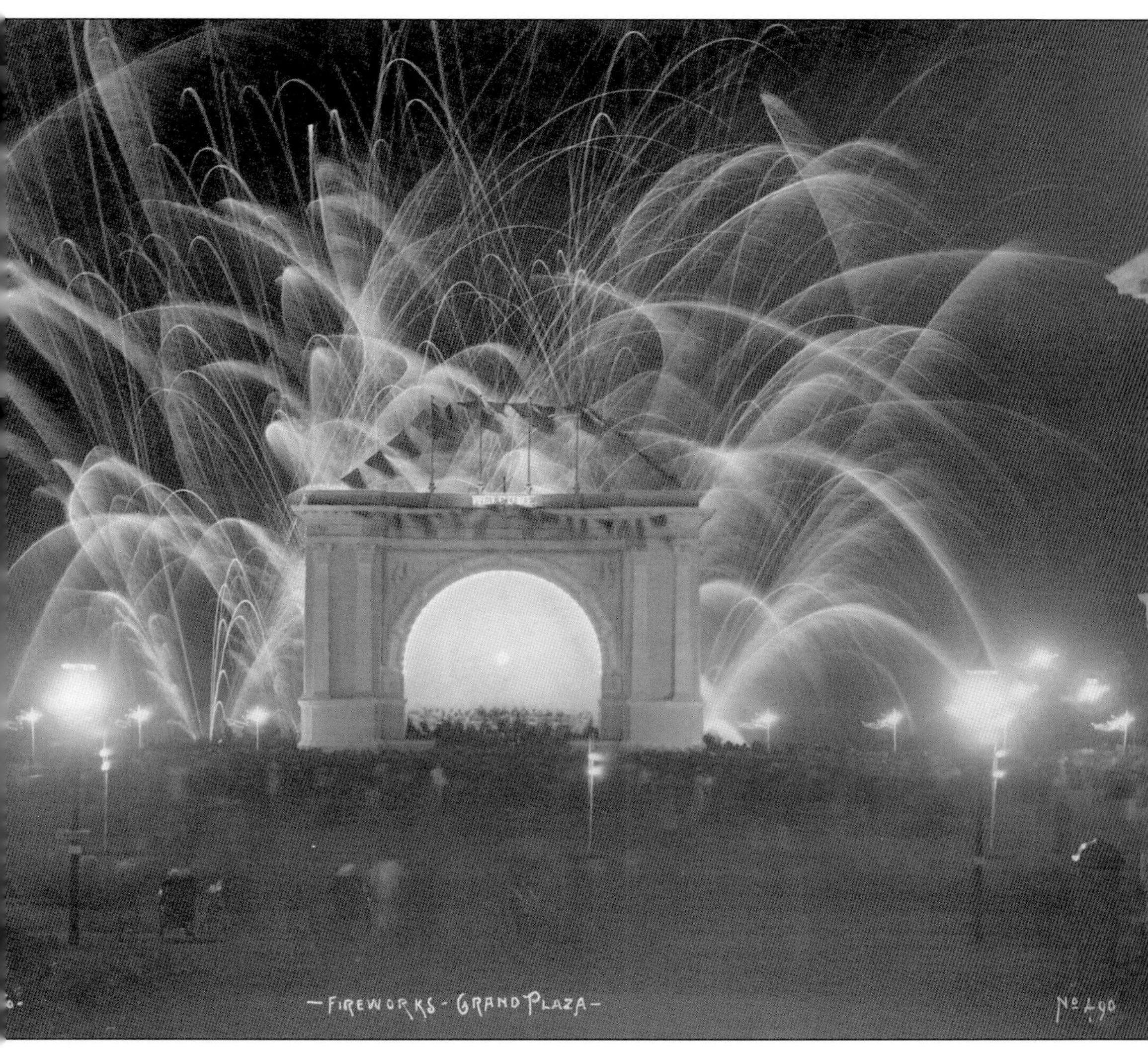

Displays of fireworks were an anticipated aspect of many celebrations at the Trans-Mississippi Exposition. This Independence Day display, set against the Music Pavilion on the Grand Plaza, was a tribute to the nation and the events of the day, as well as the skill of official exposition photographer, Frank A. Rinehart. (Photo courtesy of Omaha Public Library.)

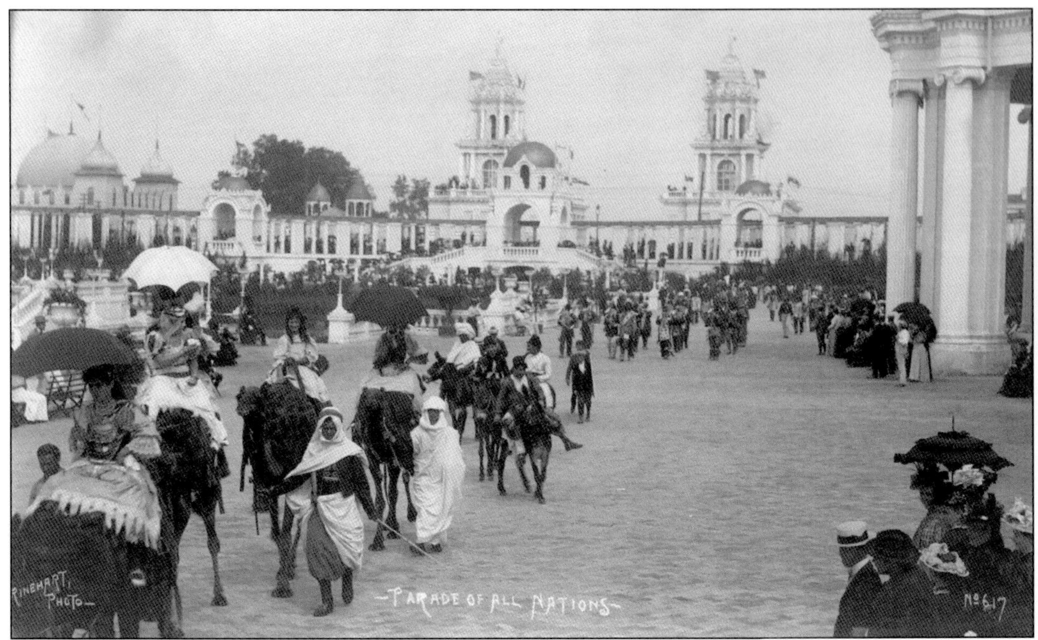

On July 4, 1898, the attractions of the midway moved to the Grand Court for the Parade of All Nations. Performers from the Streets of Cairo joined the musicians and singers from the German Village and other international performers for this event. Colorful costumes and various animals were featured in this event, welcoming 44,452 visitors to the exposition on Independence Day. (Photo courtesy of Omaha Public Library.)

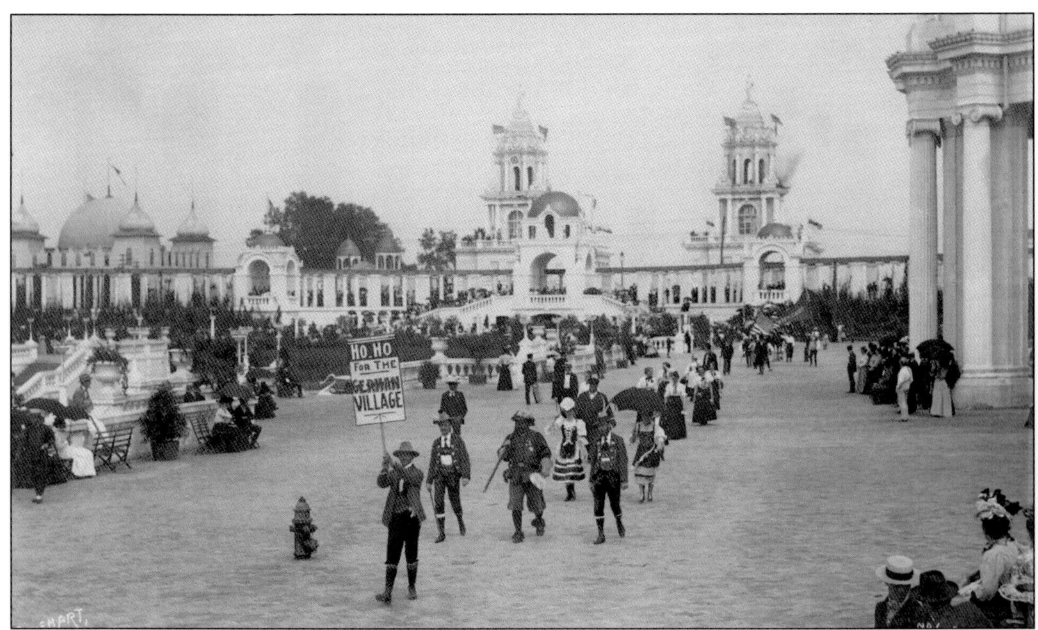

Costumed performers from the German Village march in the Parade of All Nations, led by a sign-bearer, whose placard read, "Ho Ho for the German Village." A float from Hagenback's Wild Animal Show, featuring live animals, was also included in the parade. (Photo courtesy of Omaha Public Library.)

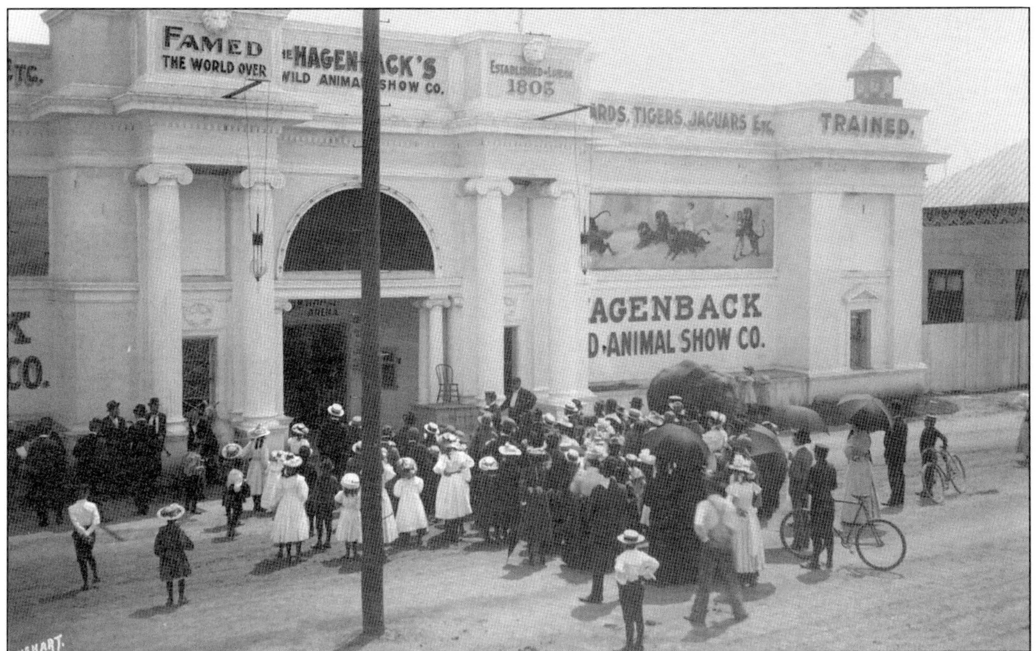

Hagenback's Wild Animal Show was a favorite attraction on Children's Day, July 14, 1898. Approximately 500 species of animals were featured in this attraction of the North Midway. This traveling show presented the talents of 20 animal trainers with nearly continuous shows. (Photo courtesy of Omaha Public Library.)

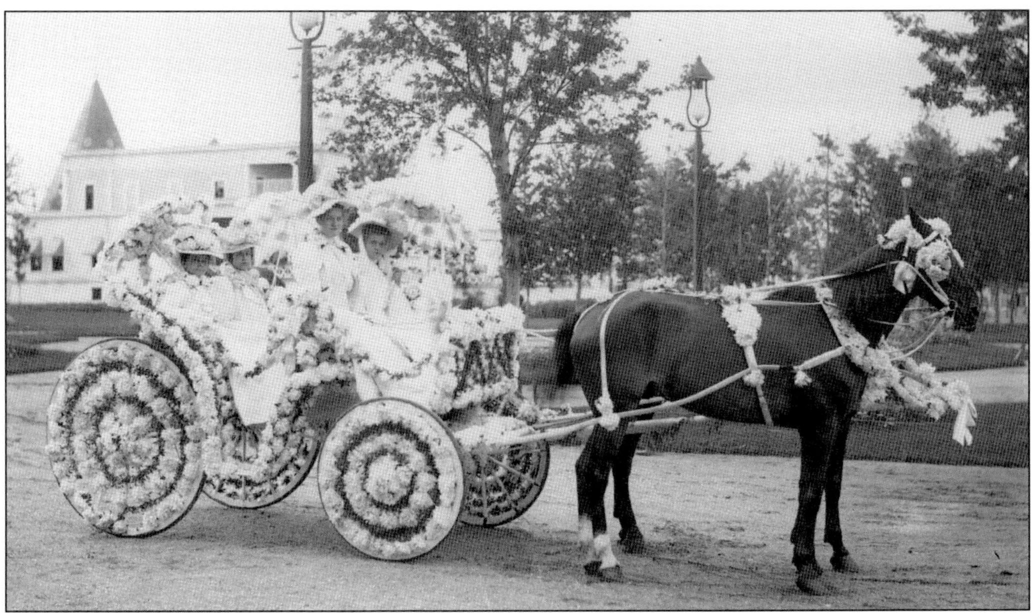

Delayed for one day due to bad weather, the highlight of the Flower Day celebration at the exposition, August 5, 1898, was the Flower Parade. Sponsored by the Ladies' Bureau of Entertainment, the Flower Parade included many carriages decorated with a variety of colorful flowers. This carriage, ready for the parade, waits near the Grand Plaza of the exposition grounds. (Photo courtesy of Omaha Public Library.)

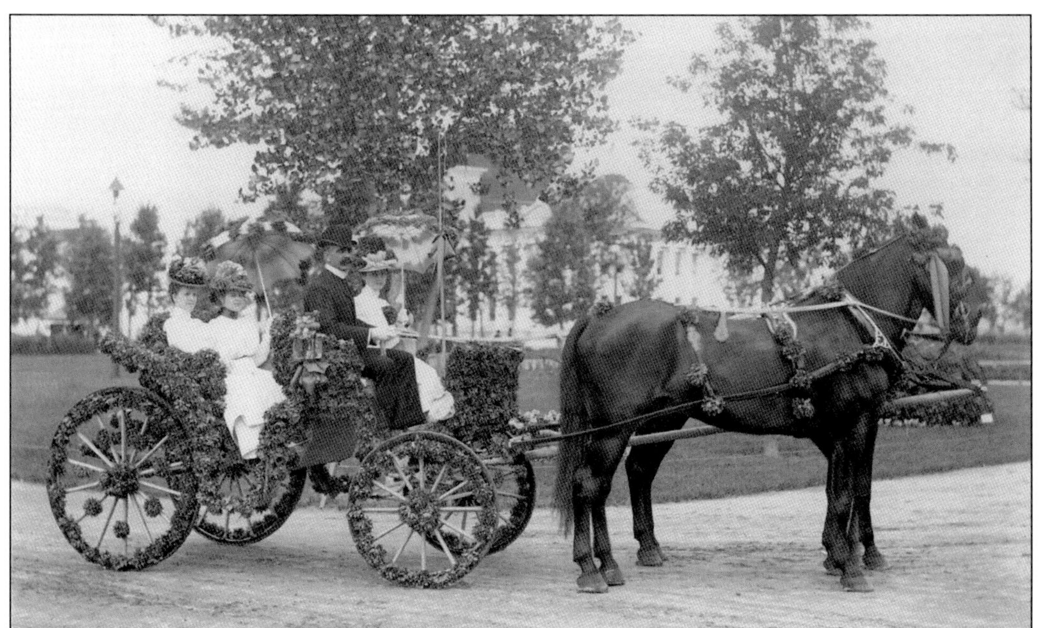

Forty decorated carriages participated in the Flower Day parade. The carriages were judged by Mayor Frank E. Moores of Omaha, Mayor Frank A. Graham of Lincoln, and Mayor Victor Jennings of Council Bluffs. All participants were awarded a souvenir medal, while first, second, and third prizes were awarded to Mrs. J.H. Evans, Mrs. Howard Baldridge, and Mrs. John N. Baldwin, respectively. (Photo courtesy of Omaha Public Library.)

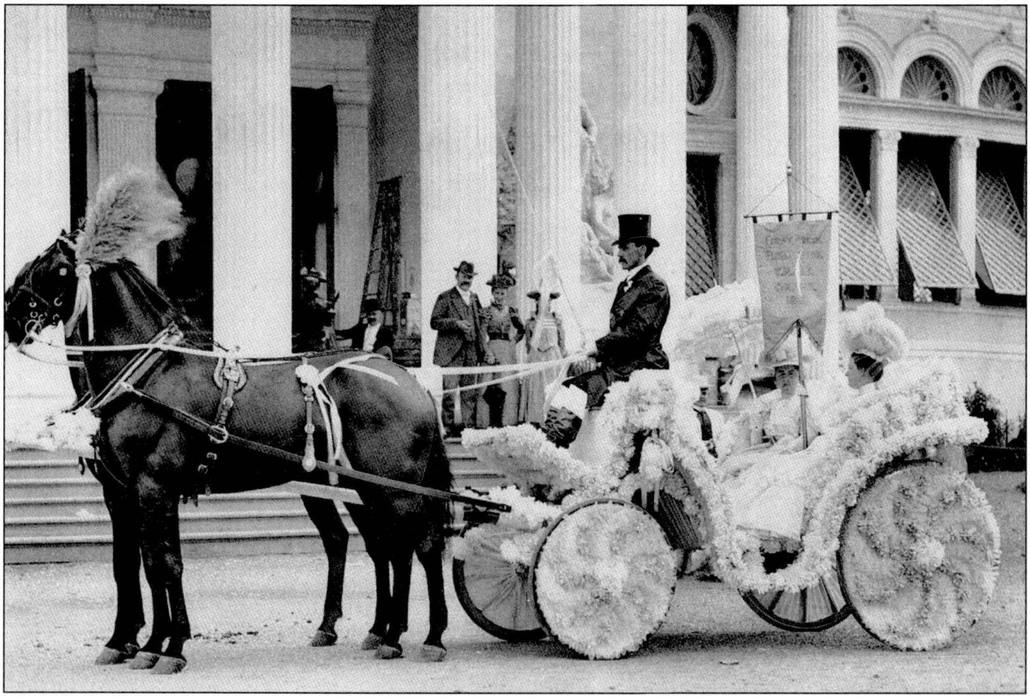

In a carriage decorated with white chrysanthemums, Mrs. J.H. Evans and Miss Amy Barker ride in their first place-winning entry in the Flower Day parade.

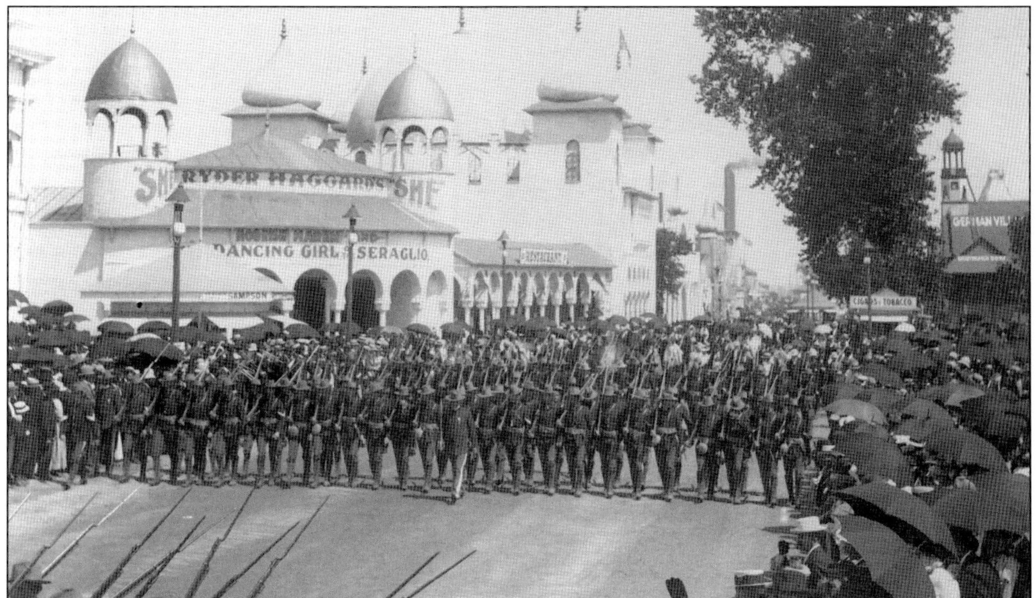

With standing orders for deployment to the Spanish-American war theater, the Third Regiment of Nebraska marched along the East Midway. Approximately 1,500 soldiers participated in Military Day exercises on July 16, 1898. The 10:00 a.m. parade circled the exposition Midway and ended when all participants reached the Grand Court. (Photo courtesy of Omaha Public Library.)

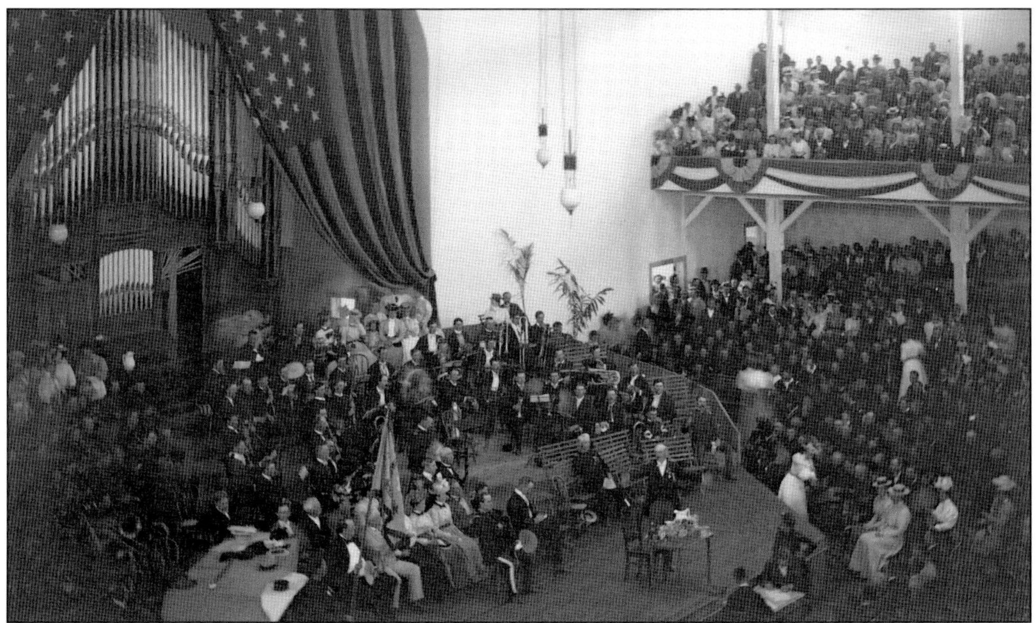

With a seating capacity of 4,500, the auditorium was the scene of many significant events during the exposition season, including this speech by Colonel William Jennings Bryan on Military Day. The three-manual organ pictured at the left was installed by Moeller of Hagerstown, Maryland. The organ was used during religious services held each Sunday during the exposition. (Photo courtesy of Omaha Public Library.)

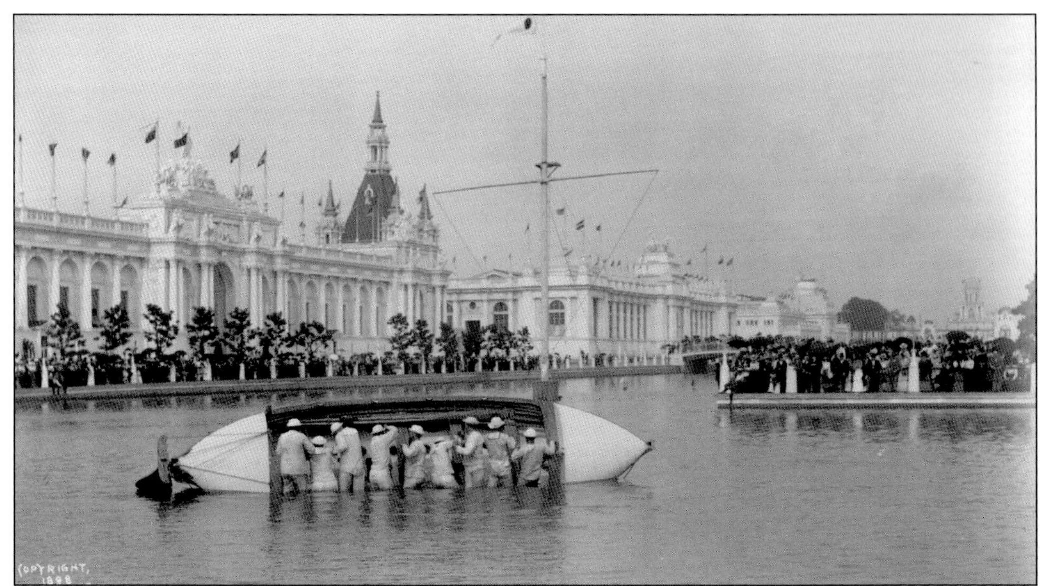

Beginning on United States Life-Saving Service Day, August 11, 1898, service members provided daily demonstrations of their technique. The demonstrations, which occurred at the mirror basin in front of the United States Government Building, included demonstrations of life boat rescues at sea and life-saving techniques for drowning victims. (Photo courtesy of Omaha Public Library.)

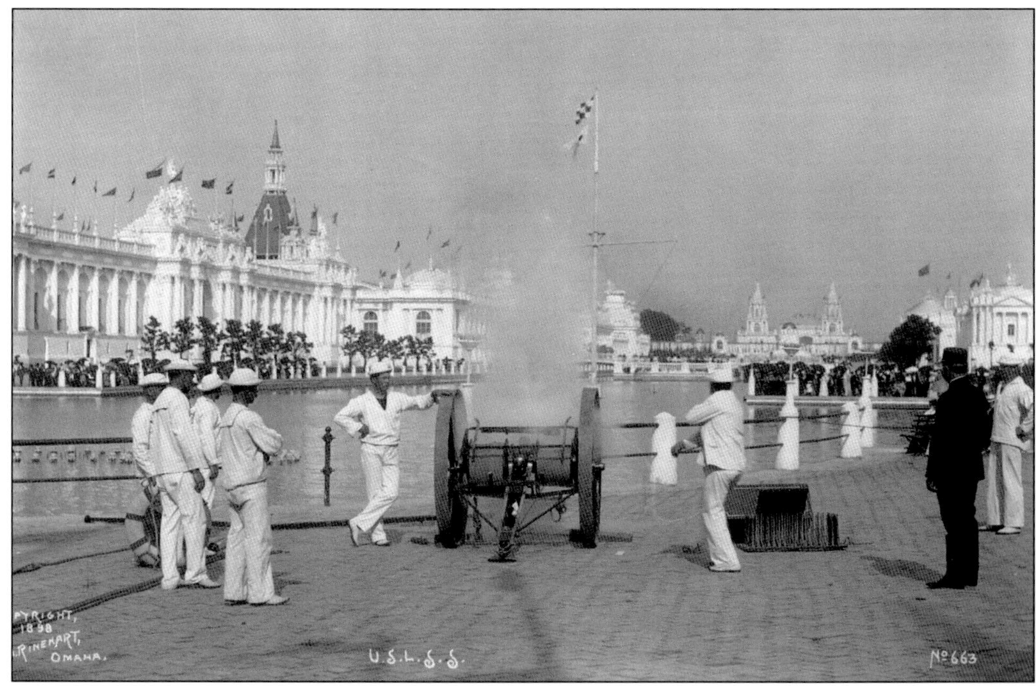

Members of the United States Life-Saving Service demonstrate the manner in which a life-saving line is fired across a distressed craft. The popularity of these demonstrations, which occurred daily at 4:00 p.m., is evidenced by the visitors in the distance who line the banks of the mirror basin. (Photo courtesy of Omaha Public Library.)

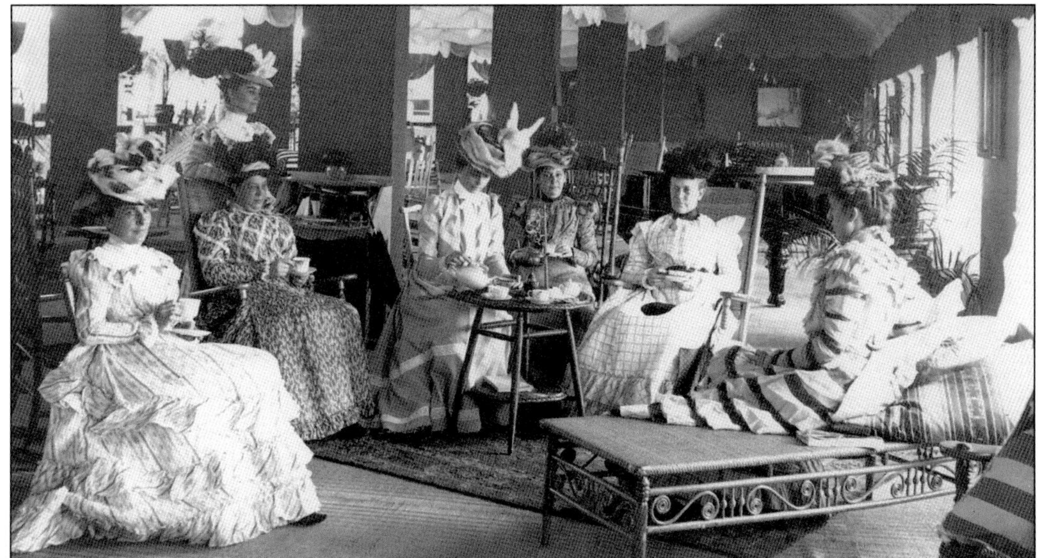

Approximately 50 local ladies comprised the Ladies' Bureau of Entertainment for the Trans-Mississippi Exposition. Ensuring a positive impression of Omaha and providing for the entertainment of visiting dignitaries were the principal tasks of this group. This space, in the second-story gallery of the Mines and Mining Building, served as the Ladies' Drawing Room. Green draperies, wicker furniture, and practical floor coverings added to this comfortable setting for receptions and relaxation for busy exposition sponsors. (Photo courtesy of Omaha Public Library.)

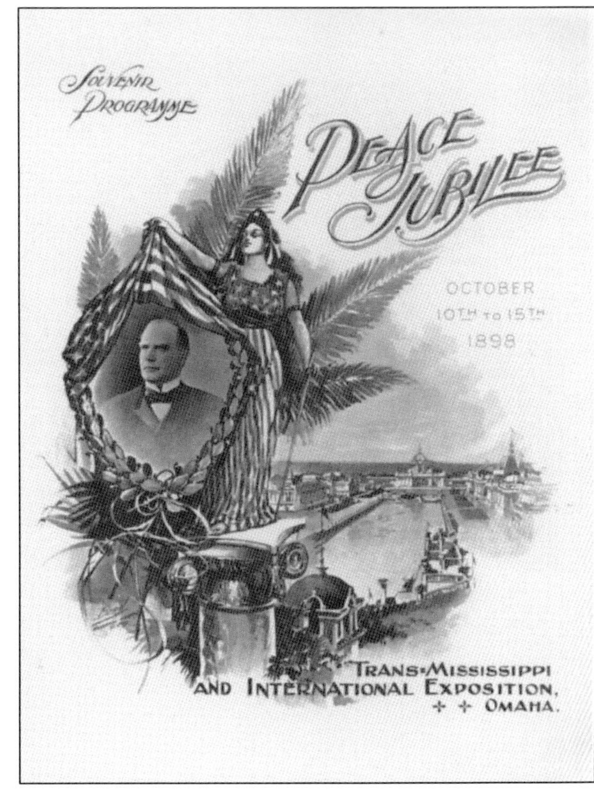

The official program of Peace Jubilee Week featured this artist's rendition of the Grand Court with a portrait of President William McKinley. Peace Jubilee Week, October 10–15, 1898, featured six "Special Days": Mayors' Day, Governors' Day, President's Day, Army and Navy Day, Civil Government Day, and Children's Day. (Photo courtesy of Omaha Public Library.)

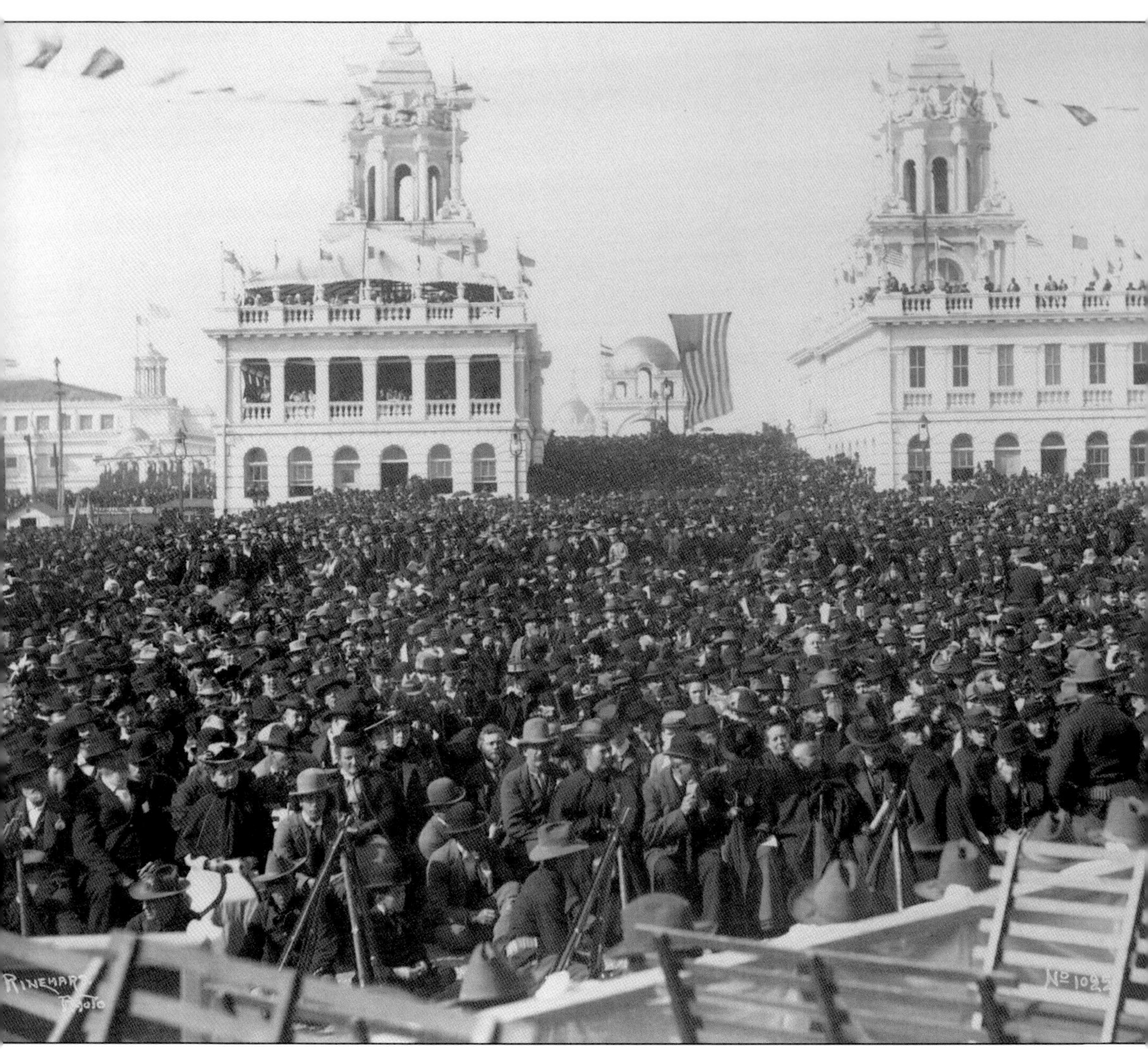

An estimated 70,000 people filled the Grand Plaza and the surrounding area in anticipation of President McKinley's arrival. A total of 98,845 people, a single-day record, visited the exposition on President's Day, October 12, 1898. In celebration of Peace Jubilee Week, which included the Presidential visit, Trans-Mississippi Exposition President Gurdon W. Wattles introduced President McKinley by stating that "[a]t no more fitting place than here in the center of our territory, surrounded by such magnificent evidences of the arts of peace, could this celebration be held." (Photo courtesy of Omaha Public Library.)

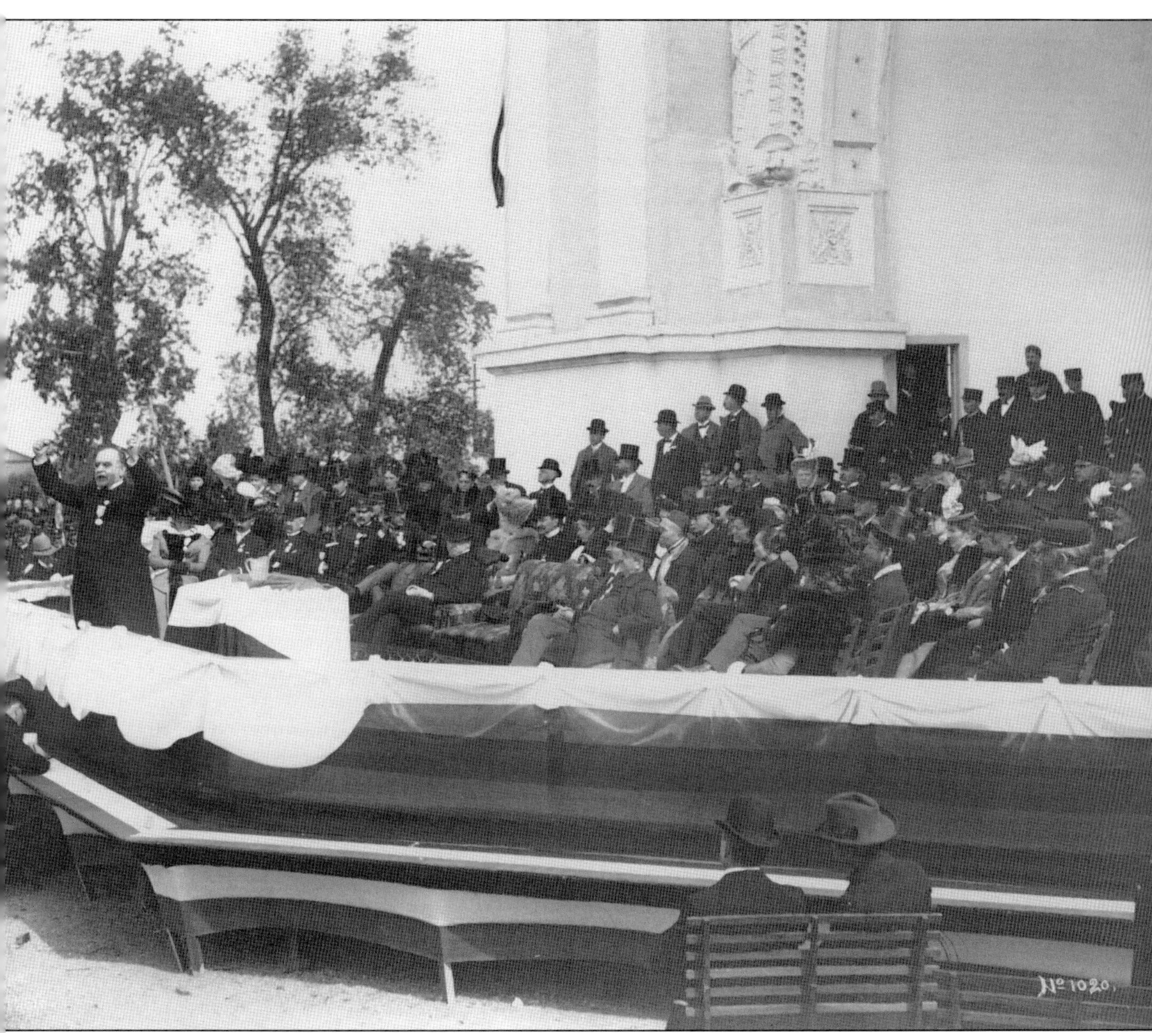

President William McKinley, from the extended stage of the Music Pavilion, congratulated the organizers of the Trans-Mississippi Exposition by exclaiming, "One of the great laws of life is progress, and nowhere have the principles of this law been so strikingly illustrated as in the United States." Reminded of the peace that had been won recently, McKinley concluded his speech with the belief that "[t]he genius of the nation, its freedom, its wisdom, its humanity, its courage, its justice, favored by Divine Providence, will make it equal to every task, and the master of every emergency." (Photo courtesy of Omaha Public Library.)

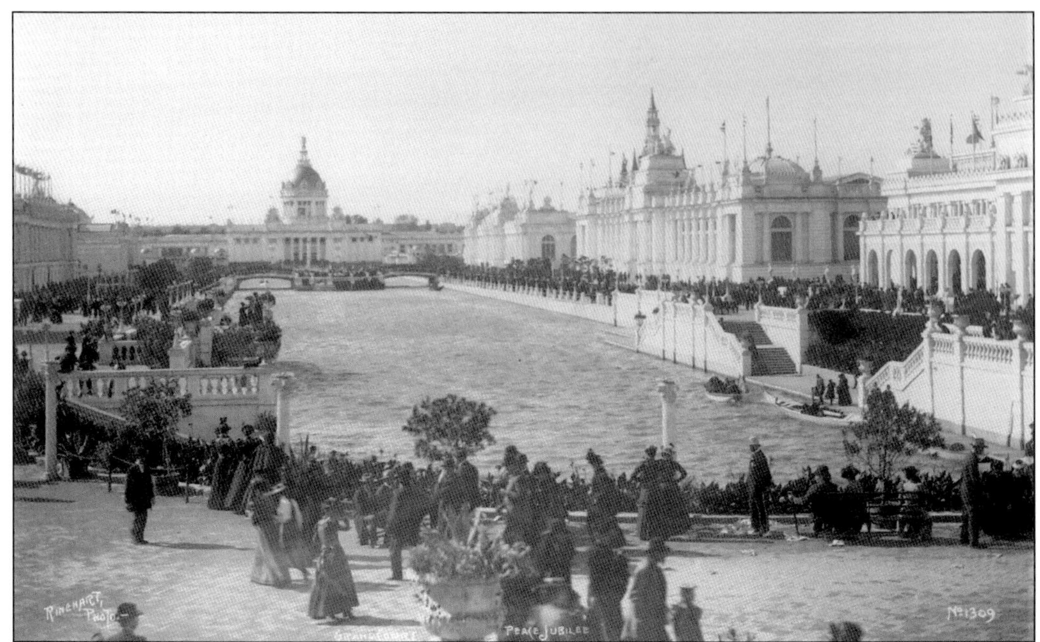

After lunch, President McKinley was joined by Trans-Mississippi Exposition President Gurdon W. Wattles on a walk through the center aisles of each building along the Grand Court. The President offered personal greetings to select visitors during a brief stopover at the United States Government Building. (Photo courtesy of Omaha Public Library.)

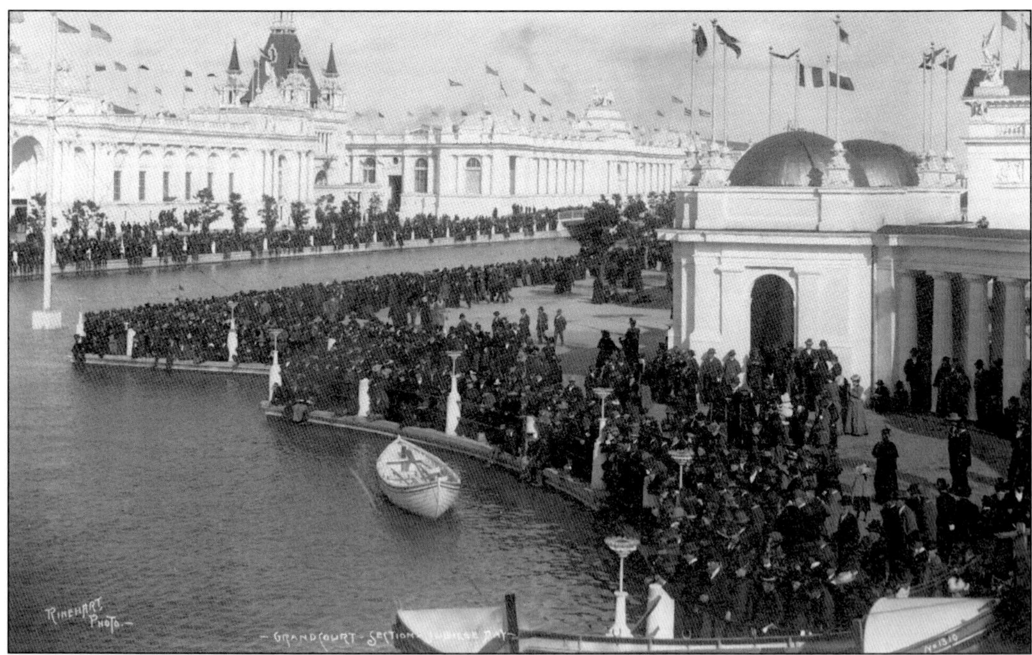

Hoping to get a glimpse of the President, exposition visitors lined the Grand Court on President's Day. Before leaving Omaha, McKinley said, "My visit to Omaha and to the Trans-Mississippi is one that I shall long remember with the kindliest recollections." (Photo courtesy of Omaha Public Library.)

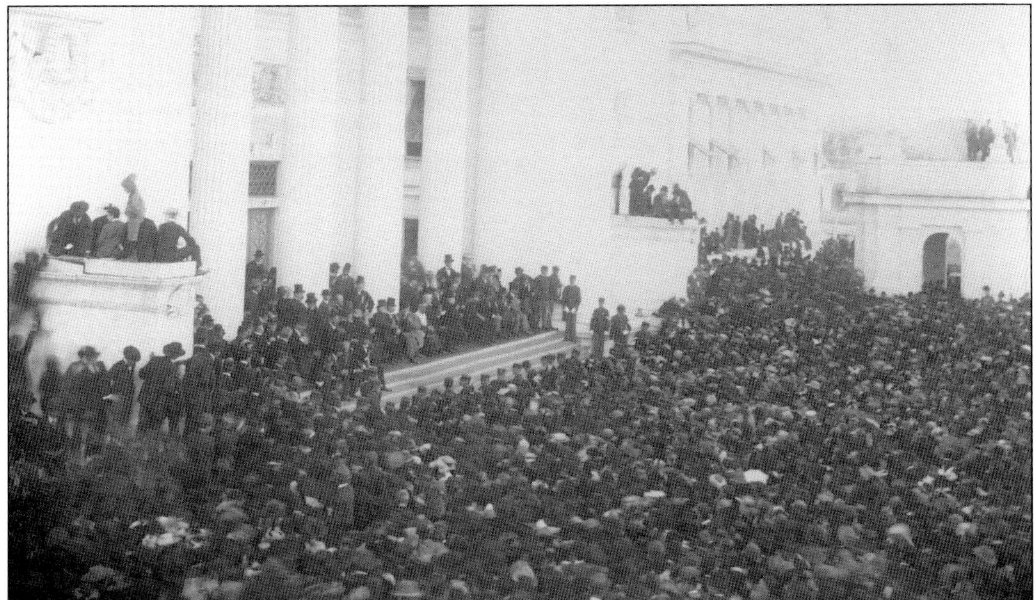

President McKinley intended to hold a public reception within the United States Government Building. The President later abandoned this idea, as such individual contact was impractical, given the large crowd. Perhaps McKinley's decision was wise, as he would fall victim to an assassin's gun three years later at the 1901 Buffalo Pan-American Exposition. (Photo courtesy of Omaha Public Library.)

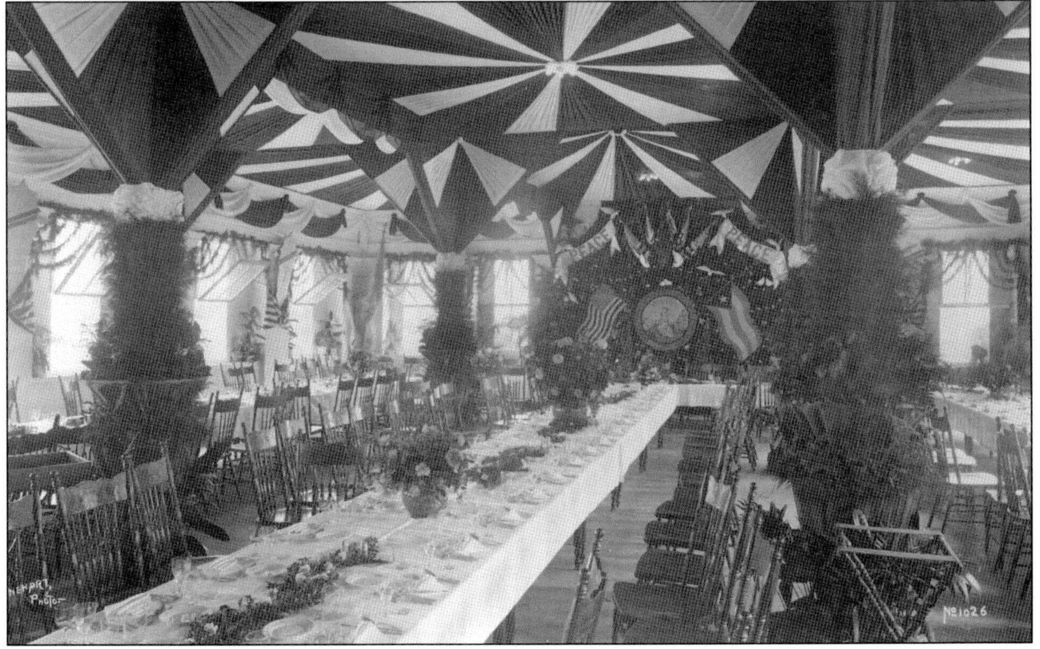

The Markel Café within the twin restaurants at the South Viaduct of the exposition was the location for the "Peace Jubilee Dinner" given in honor of President McKinley's visit. Dinner entrées included planked whitefish with fine herbs, dressed cucumbers, braised lamb chops and roast canvasback duck with cresses. (Photo courtesy of Omaha Public Library.)

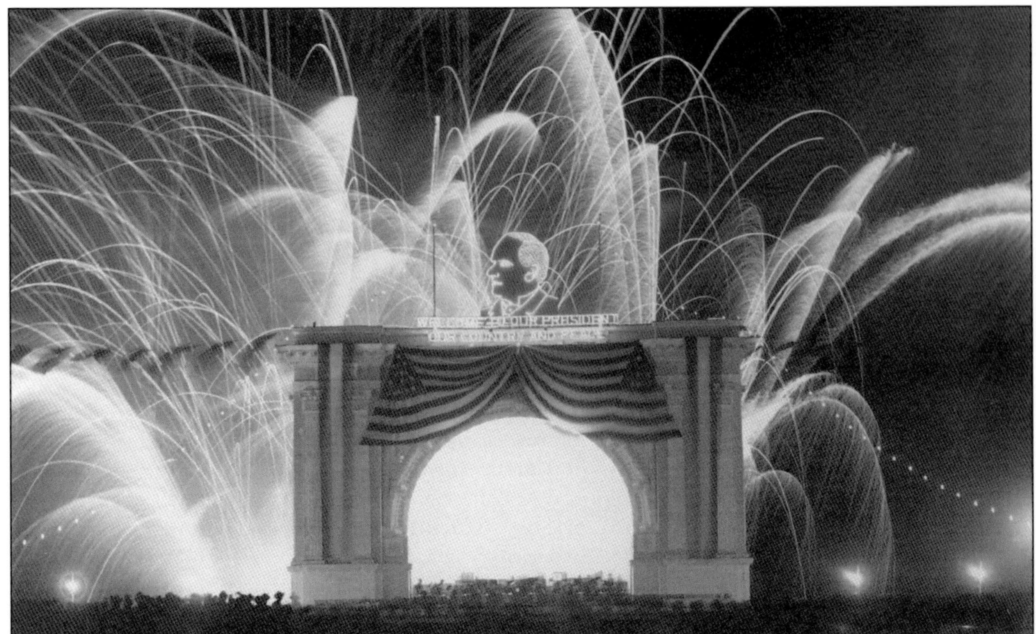

After a formal dinner, President McKinley enjoyed a spectacular program of music and a display of fireworks over the Grand Plaza, the site of his morning speech. Thousands of exposition visitors remained on the grounds throughout the day in anticipation of the evening celebration. (Photo courtesy of Omaha Public Library.)

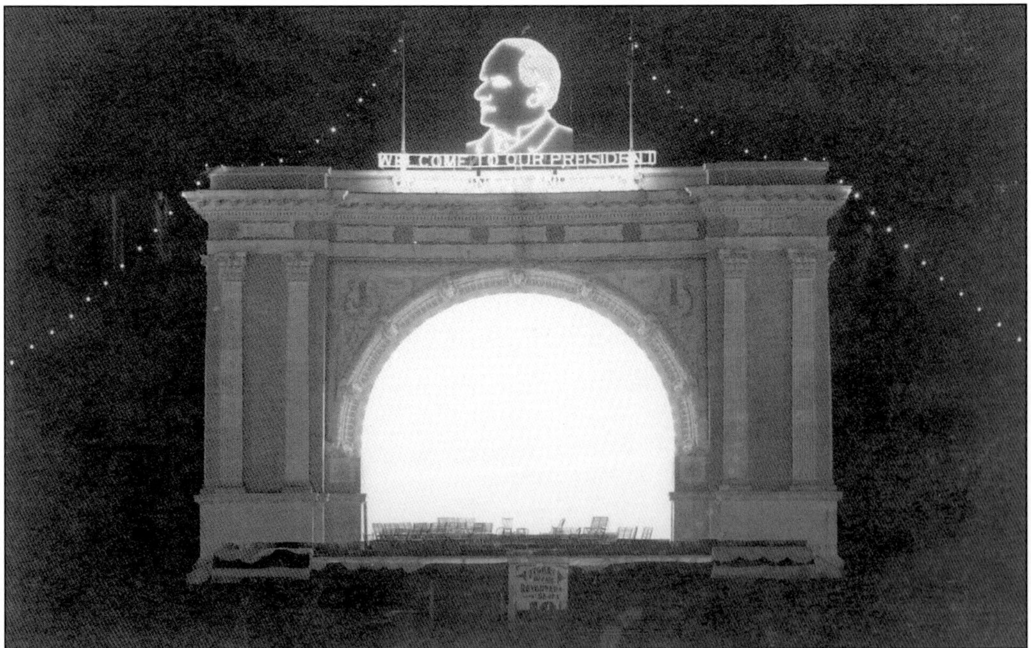

After the celebration ended, visitors left the Grand Plaza along with President McKinley. Official exposition photographer Frank A. Rinehart remained for one last picture. An illuminated bust of the President above the words "Welcome to Our President, Our Country and Peace" is pictured. (Photo courtesy of Omaha Public Library.)

Five

A Proud Heritage Preserved

Regarded by the *Omaha Daily Bee* as "one of the most unique and attractive features of the exposition," the concurrent Indian Congress was among the most historically significant events of the Trans-Mississippi Exposition season. While historical study has questioned the contention, according to the official history of this grand event, that "this unique exhibit enabled thousands of visitors to see what the government had done for the benefit and welfare of the aborigines," the photographs and recorded events of the congress are treasures of the American West.

Situated on the North tract of the exposition grounds, the four-acre Indian Encampment featured Native Americans from an estimated 35 distinct tribes throughout the United States, most of which had been relocated to reservations. Native American representatives were encouraged to showcase their history with traditional dress and other practices intended to exhibit the bygone era of a disappearing culture.

A plan to present a more studied exhibit of Native American life was, in many ways, discarded for showmanship, featuring sham battles, dances, and other events that added to the spectacle of the Indian Congress. Visitors who looked beyond the visual excitement of these events, however, discovered Native Americans with rich, diverse traditions.

Progress in the Trans-Mississippi region was, in part, earned at the cost of displaced Native Americans. For a culture based upon an oral tradition, the photographs of the Indian Congress are a time-honored resource into the daily life of their ancestors, the spirits of whom embrace modern generations.

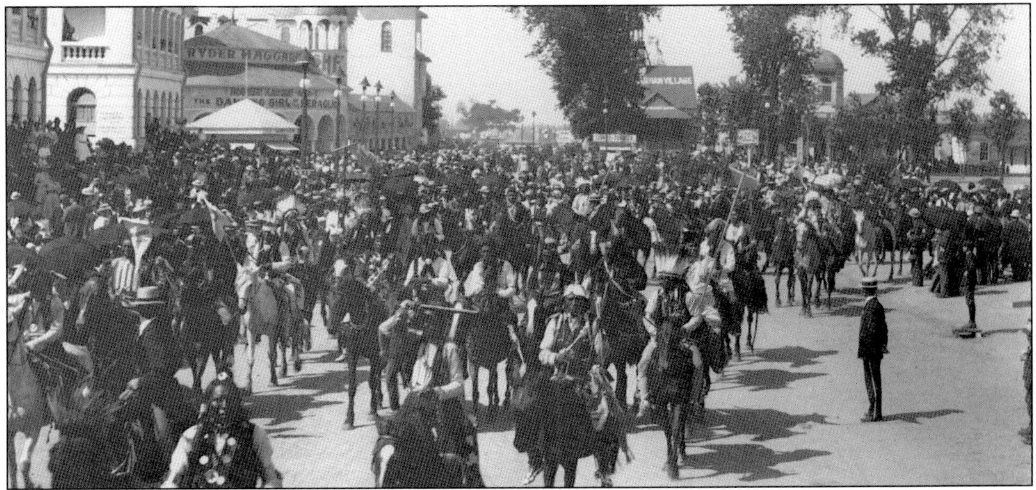

August 4, 1898, was designated as Indian Day at the Trans-Mississippi Exposition. The parade of Native Americans was a key event fulfilling the desire of exposition planners to present the culture and traditions of various tribes. Pictured here is the procession as it passes through the Grand Plaza. Approximately 150 Native Americans later participated in a parade through the streets of downtown Omaha. (Photo courtesy of Omaha Public Library.)

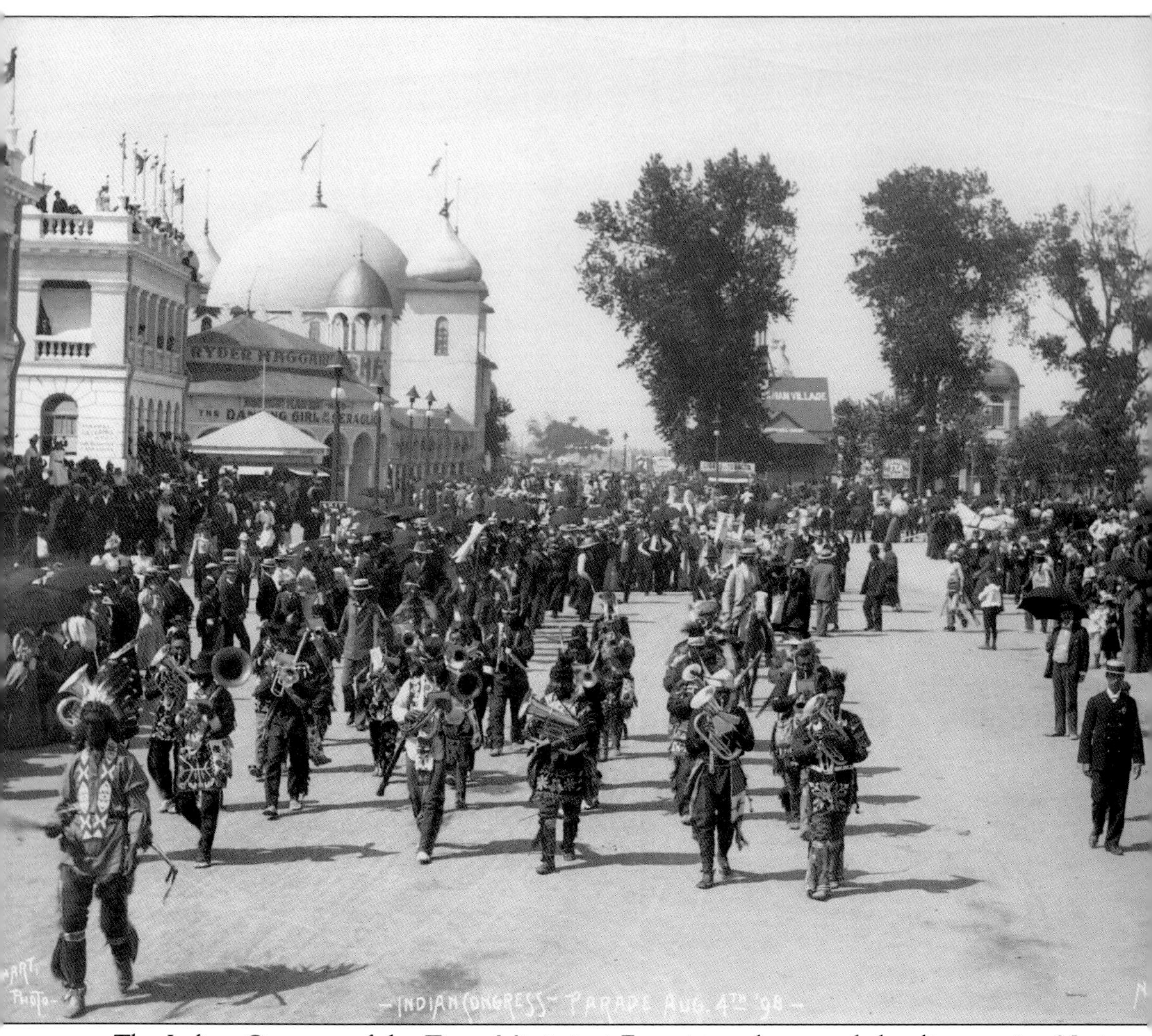

The Indian Congress of the Trans-Mississippi Exposition showcased the disappearing Native American way of life. The Indian Day Parade featured representatives from many tribes in traditional and ceremonial dress. The parade, which officially opened the Indian Congress, was followed by celebrations, traditional dances, and equestrian demonstrations. Witnessing these proud traditions was a noted attraction for many visitors, previously unfamiliar with Native American culture. (Photo courtesy of Omaha Public Library.)

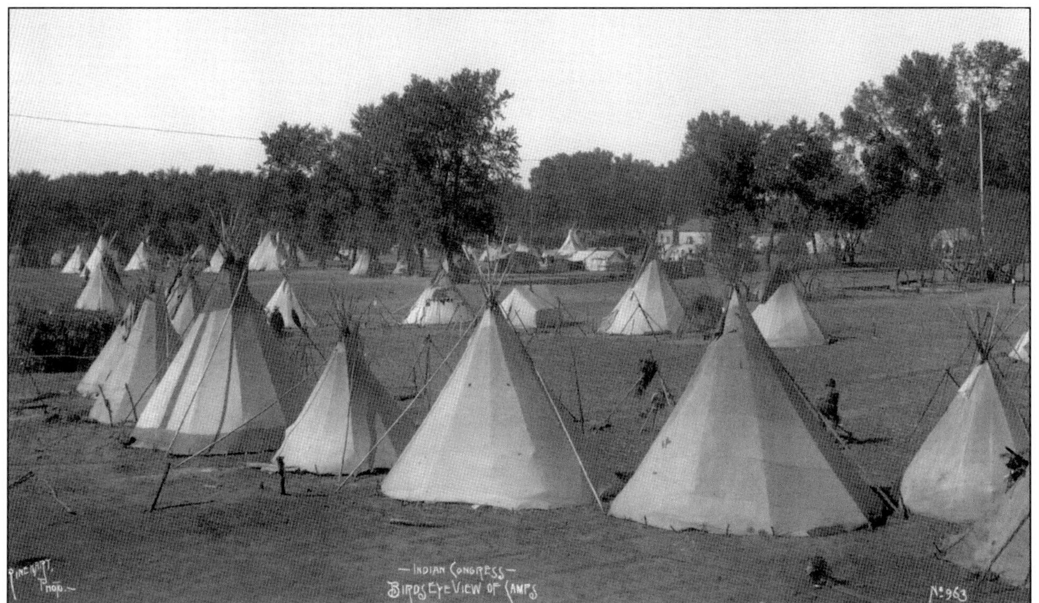

It was estimated that up to 500 Native Americans remained in the encampment during the three months of the Indian Congress. While some tribes were represented for only a short time, tribes whose members remained for the duration of the congress included those of the Apache, Arapahoe, Assiniboine, Blackfoot, Cheyenne, Crow, Flathead, Iowa, Kiowa, Omaha, Otoe, Ponca, Pottawattamie, Pueblo, Sauk and Fox, Sioux, Tonkawa, Wichita, and Winnebago. (Photo courtesy of Omaha Public Library.)

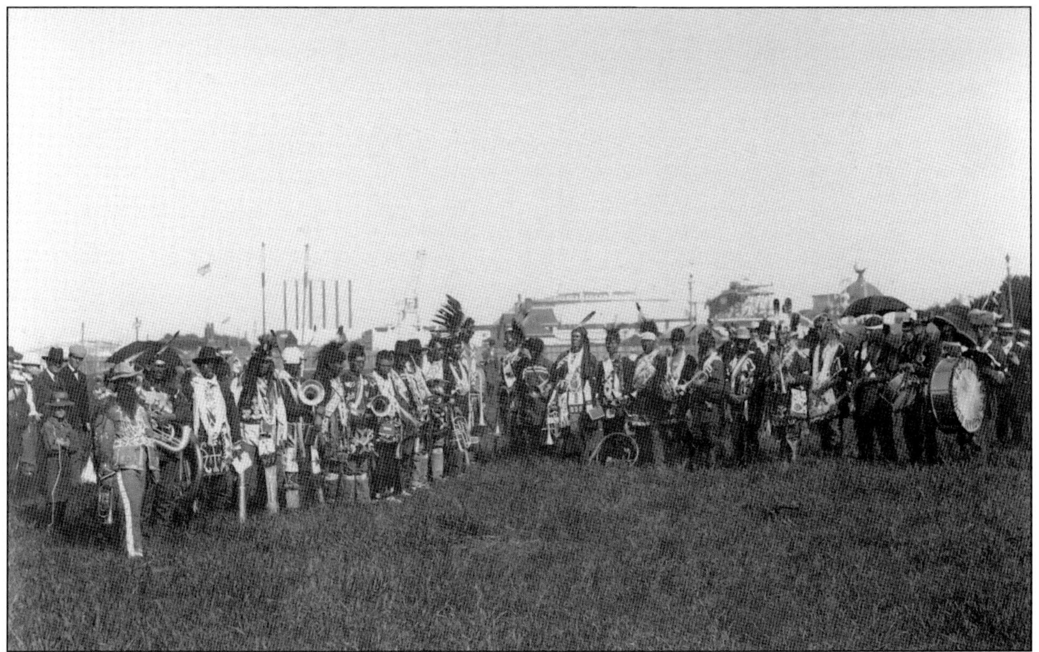

This band of Santee musicians from the U.S. Indian School of Flandreau, South Dakota, was featured among the performers at the Indian Congress. Many exposition visitors watched as this group led the parade to open the Indian Congress. (Photo courtesy of Omaha Public Library.)

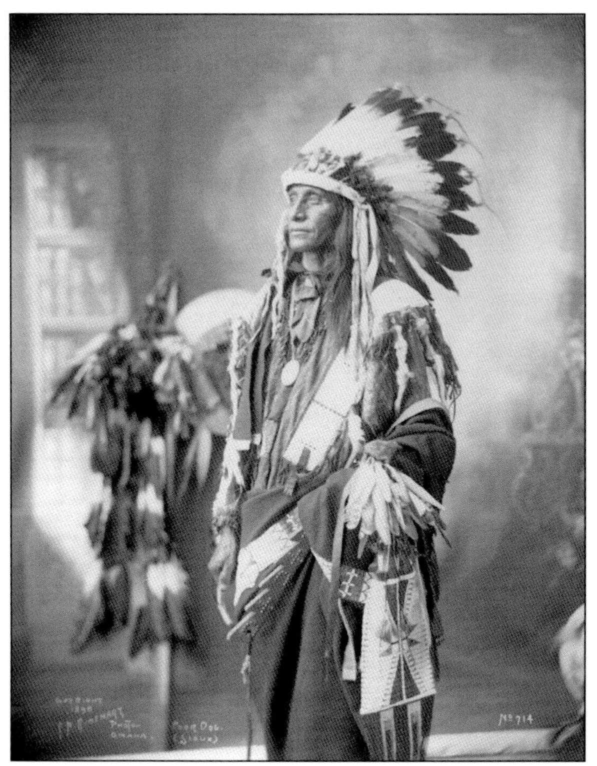

Poor Dog was among the 88 members of the Sioux tribe in attendance at the Trans-Mississippi exposition. Sioux were appreciated for their spectacular dress and, in particular, the feathered war bonnets worn by many representatives during parades and other ceremonies of the exposition season. (Photo courtesy of Omaha Public Library.)

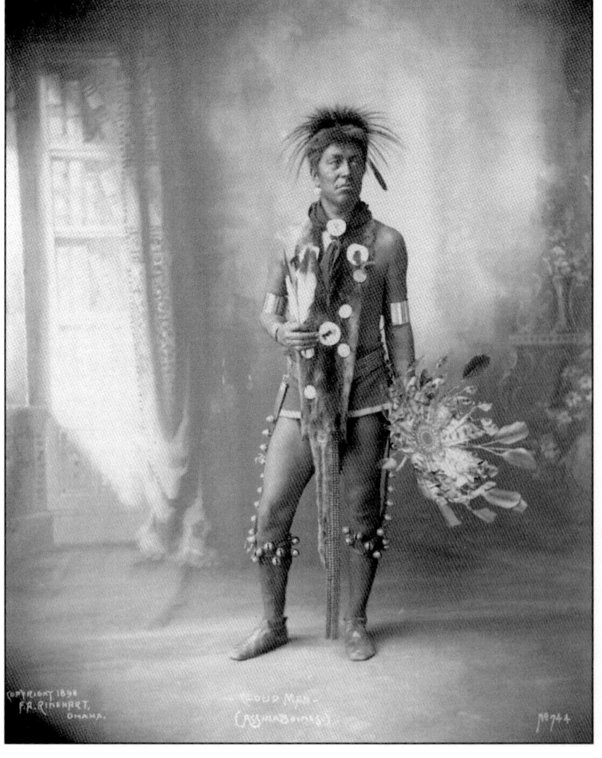

Twenty-five members of the Assiniboine tribe from Fort Peck, Montana, were represented at the exposition. Cloud Man represented his tribe, whose ancestral lands were in the Northern Plains of the United States, as well as lands across the Canadian border in the plains region. (Photo courtesy of Omaha Public Library.)

Four Bull, of the Assiniboine tribe, is featured in this standing portrait. (Photo courtesy of Omaha Public Library.)

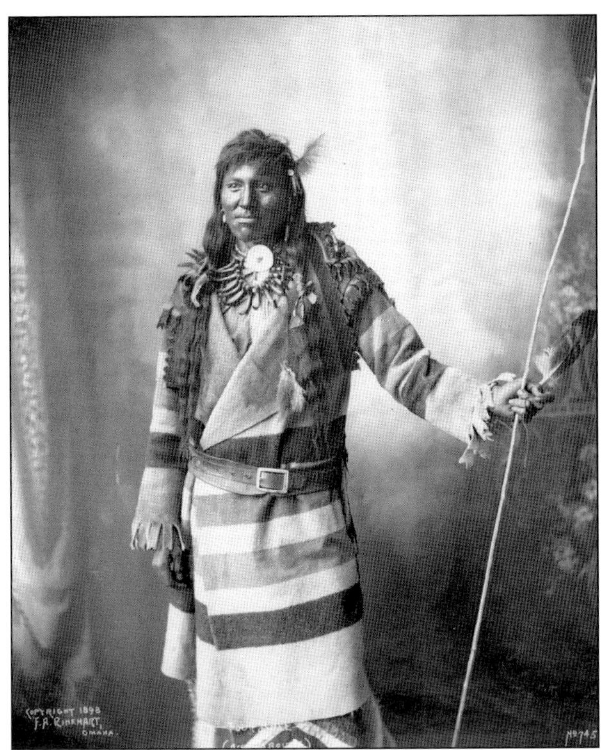

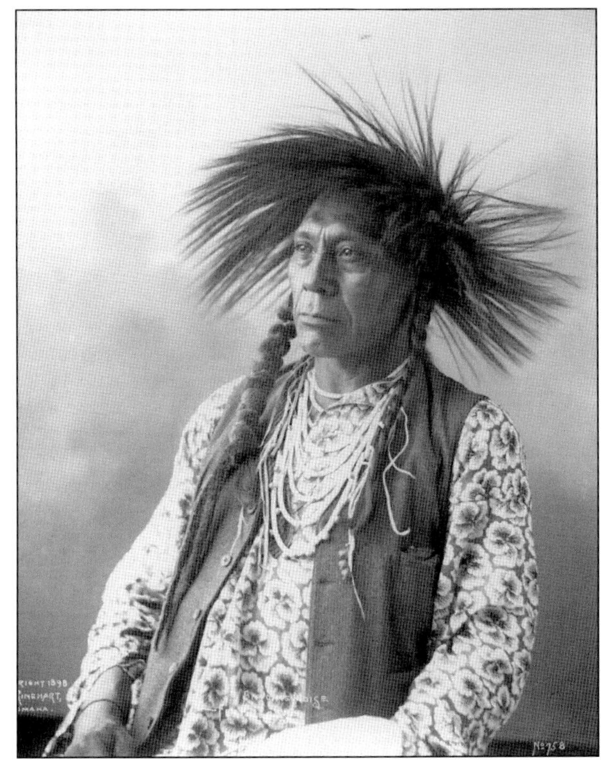

Antonnemoise, representing the Flathead tribe of Montana, displays the traditional hairstyle of tribal men in this portrait. The term Flathead did not represent the nature of tribe members' skulls but was mistakenly applied to members of the Selish, Sihqomen, and Kalispiel tribes by early visitors to the ancestral lands of these tribes. (Photo courtesy of Omaha Public Library.)

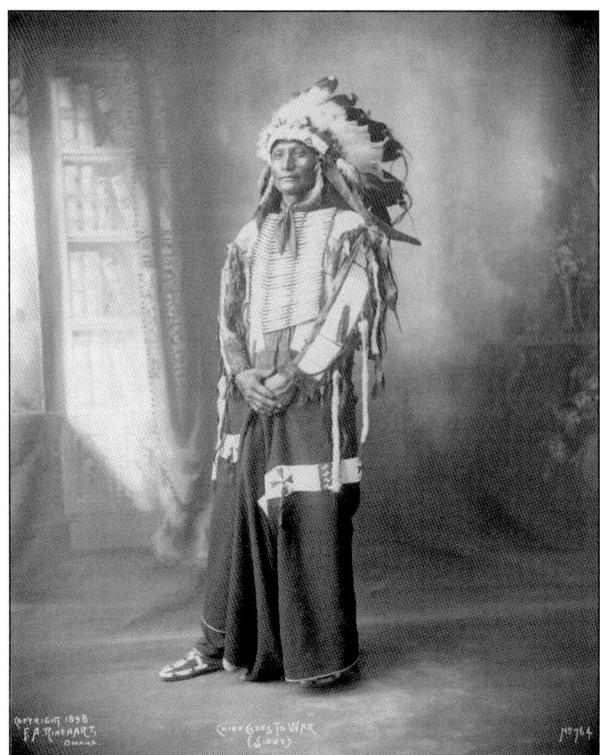

Chief Goes to War displays the historic traditions of the Sioux nation in this portrait, featuring his spectacular headdress and breast plate. The Sioux comprised the largest permanent delegation of any tribe at the Indian Congress. (Photo courtesy of Omaha Public Library.)

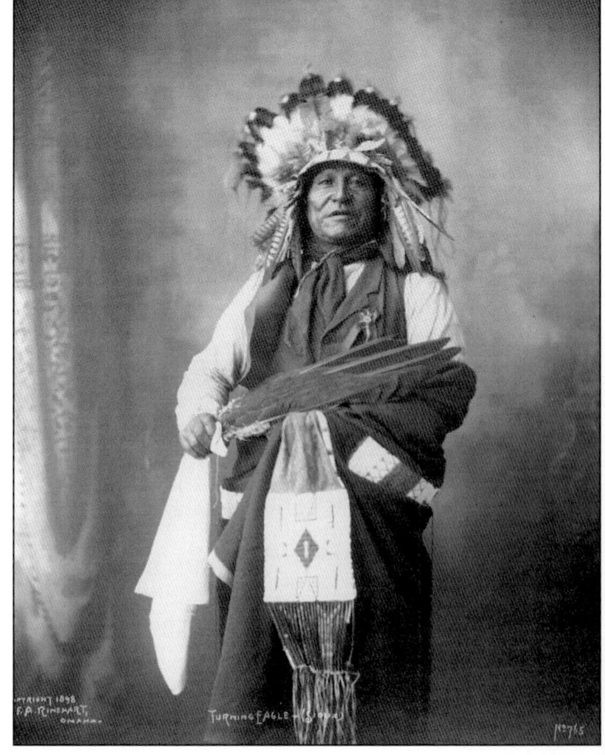

Featured in this portrait is Turning Eagle, a member of the Sioux tribe visiting the exposition. Sioux delegations traveled from South Dakota reservations in Rosebud, Crow Creek, Cheyenne River, Lower Brule, and Pine Ridge. An additional delegation represented Standing Rock, positioned on the border between North Dakota and South Dakota. (Photo courtesy of Omaha Public Library.)

Spies on the Enemy was one among 26 representatives from the Crow tribe of Montana. Ornamental dress and accessories, featured in the portrait, were a proud display for Native American attendees of the Indian Congress and a constant fascination for visitors who witnessed this very popular aspect of the Trans-Mississippi Exposition. (Photo courtesy of Omaha Public Library.)

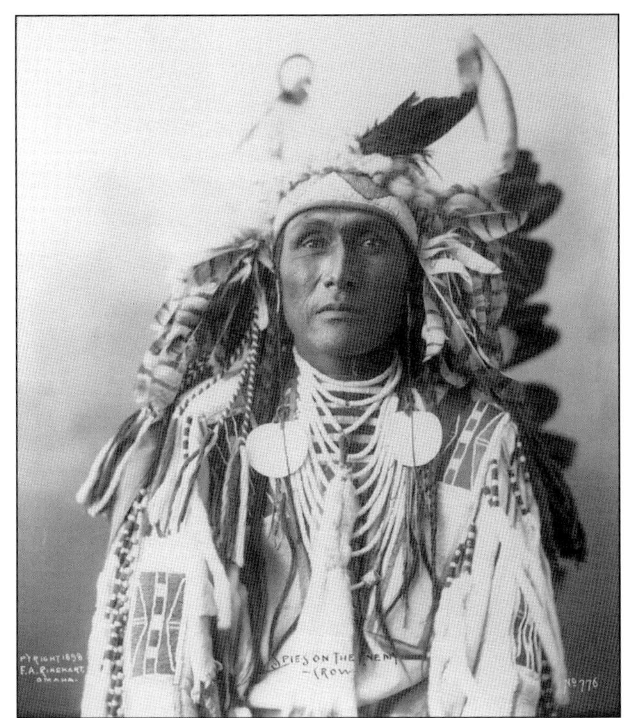

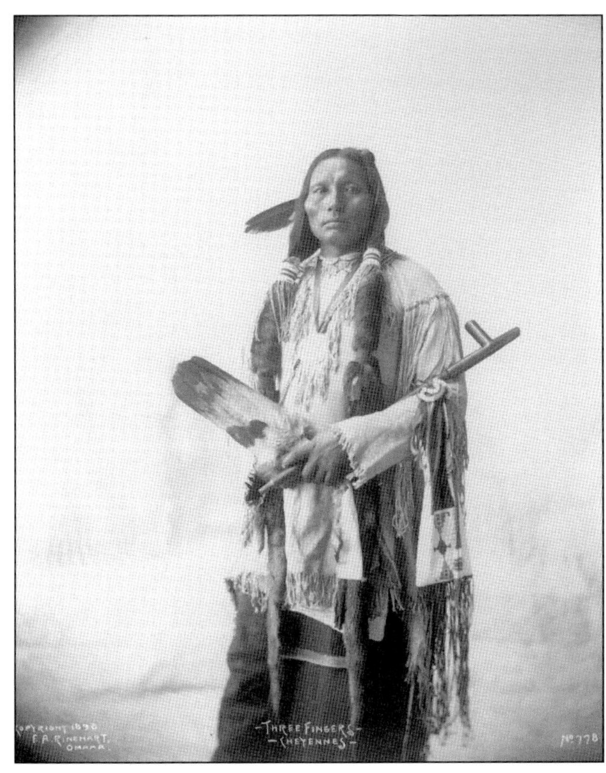

Three Fingers, of the Cheyenne tribe, clutches a ceremonial pipe and feathers in this portrait. Forty-three members of the Cheyenne nation from Oklahoma attended the Indian Congress. (Photo courtesy of Omaha Public Library.)

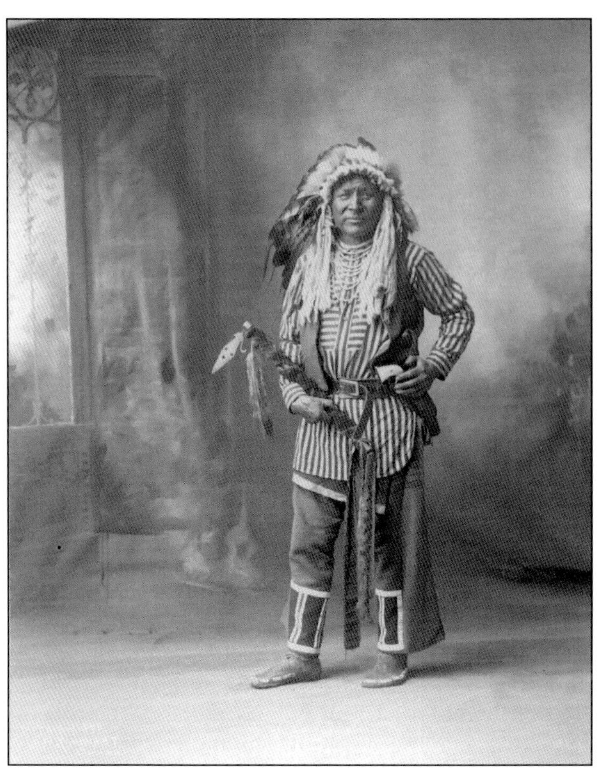

White Swan, a Crow scout noted for his service to General Custer at the Battle of the Little Bighorn in 1876, was one among many distinguished Native American exposition visitors. The opportunity to have White Swan present his story in sign language, due to a profound hearing loss resulting from battle wounds, was a prized event for those fortunate enough to have received this rare, first-hand account of American history. (Photo courtesy of Omaha Public Library.)

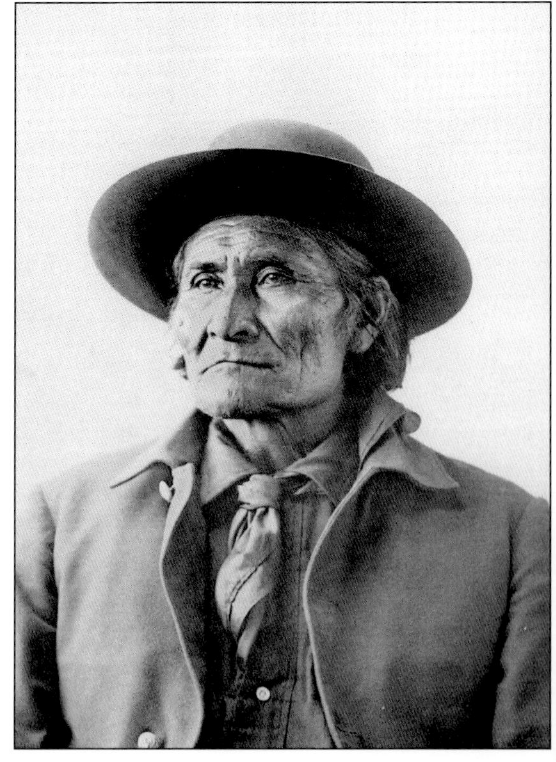

Geronimo, the famous Apache leader, visited the Trans-Mississippi Exposition as a Native American prisoner from Fort Sill, Oklahoma. In an historic embrace, Geronimo met his former captor, General Nelson A. Miles, on Army and Navy Day, October 13, 1898. This photograph is unique among the collection from the Indian Congress as Geronimo chose not to appear in traditional dress.

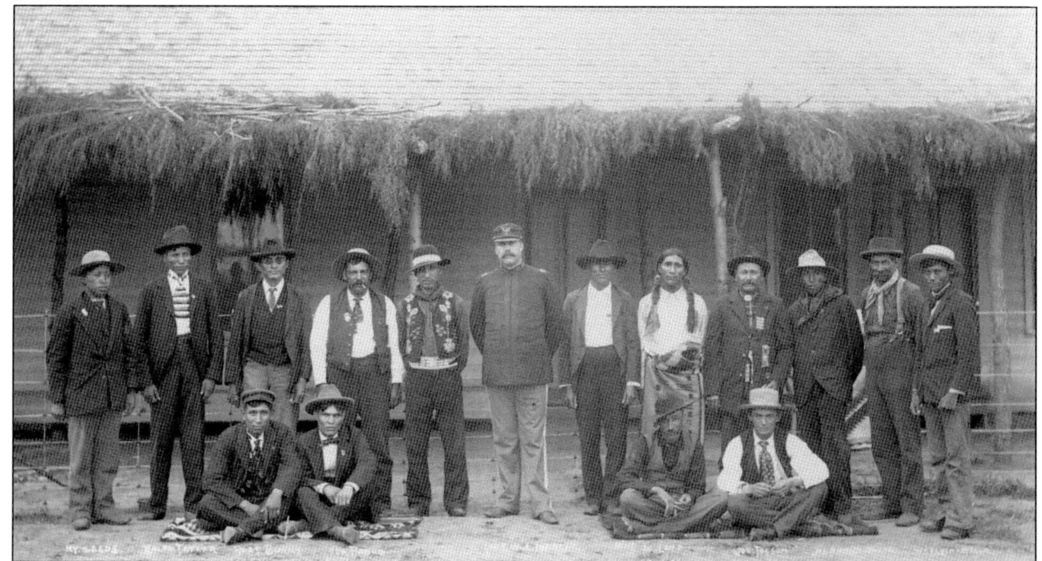

Captain W.A. Mercer of the Eighth United States Infantry and agent of the Omaha and Winnebago tribes directed the Indian Congress. In this photograph, Captain Mercer poses with his interpreters. Pictured from left to right are (front row) Joe Shooter, Moses Delaware, Dan Martin, and Jim Stevens; (back row) Hy Leeds, Ralph Taylor, Robert Burns, George Pablo, unidentified, Captain Mercer, Ben T., Ed Ladd, Joe Tesson, Joe Americanhorse, unidentified, and William Eskennatella. (Photo courtesy of Omaha Public Library.)

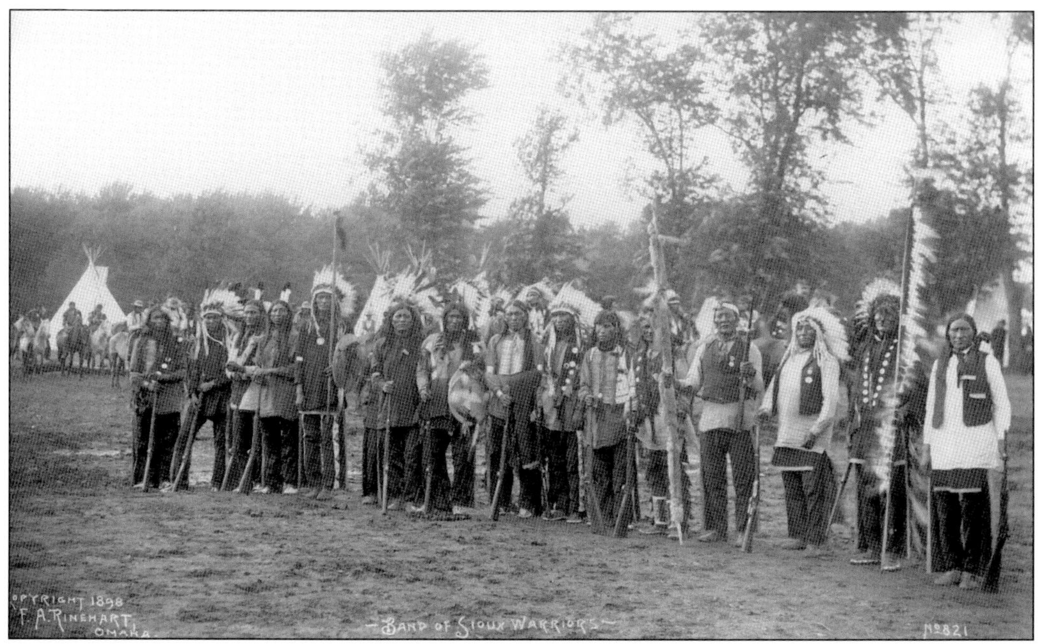

Fifteen members of the Sioux tribe, with rifles and other implements of battle, posed for this picture within the grounds of the Native American camp. To properly represent the traditional Native American lifestyle, tribal representatives were asked to bring appropriate clothing, including ceremonial dress; materials to construct, decorate, and furnish authentic shelters; and the means to assemble craft items, which could be offered for sale. (Photo courtesy of Omaha Public Library.)

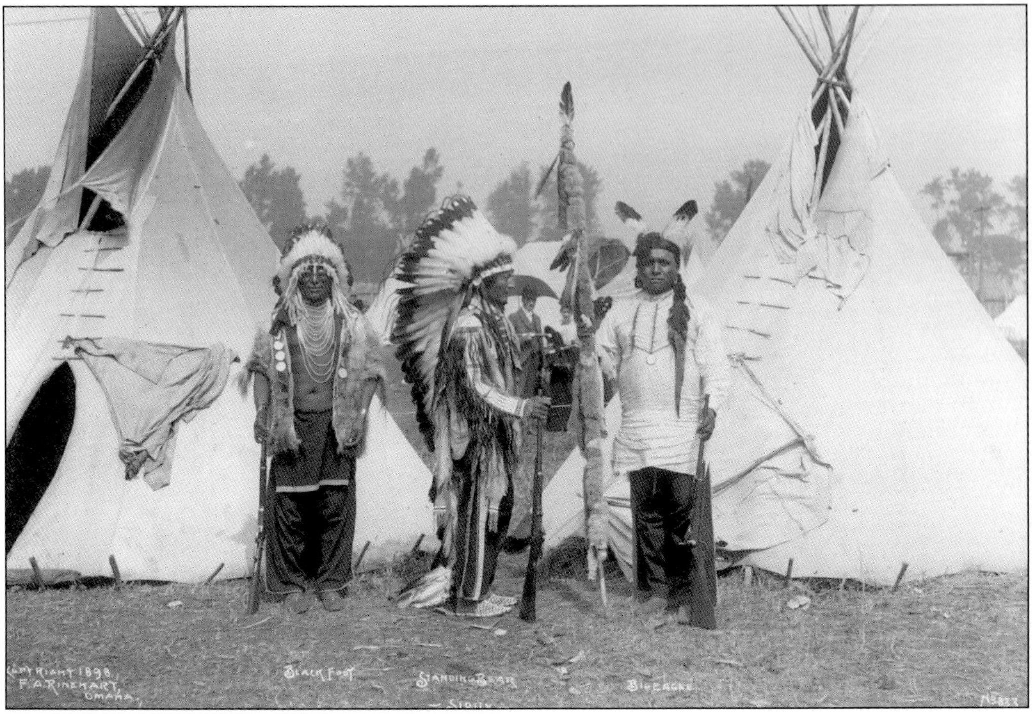

Pictured with tepees in the Indian encampment, these three Sioux representatives to the Indian Congress, from left to right, are Black Foot, Standing Bear, and Big Eagle. For 25¢, exposition visitors gained admittance to the Indian encampment on the North tract of the exposition grounds. For many visitors, this was their first, and last, opportunity to witness a live setting that purported to represent the historical traditions of Native Americans. (Photo courtesy of Omaha Public Library.)

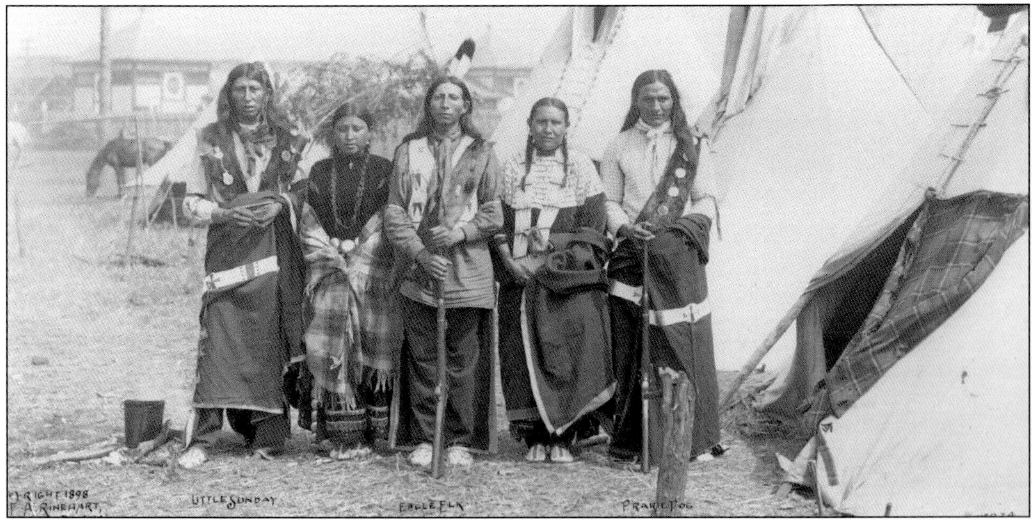

With a collection of over 500 photographs, Frank A. Rinehart, official exposition photographer, preserved memories of the Indian Congress. This group of Sioux representatives, from left to right, are, Little Sunday, unidentified, Eagle Elk, unidentified, and Prairie Dog. (Photo courtesy of Omaha Public Library.)

Three Apache representatives to the Indian Congress were identified, from left to right, simply as Bart, Tom, and Assuz. Twenty-one individuals from San Carlos, Arizona, and Chiricahua prisoners from Fort Sill, Oklahoma, comprised the Apache delegation to the congress. (Photo courtesy of Omaha Public Library.)

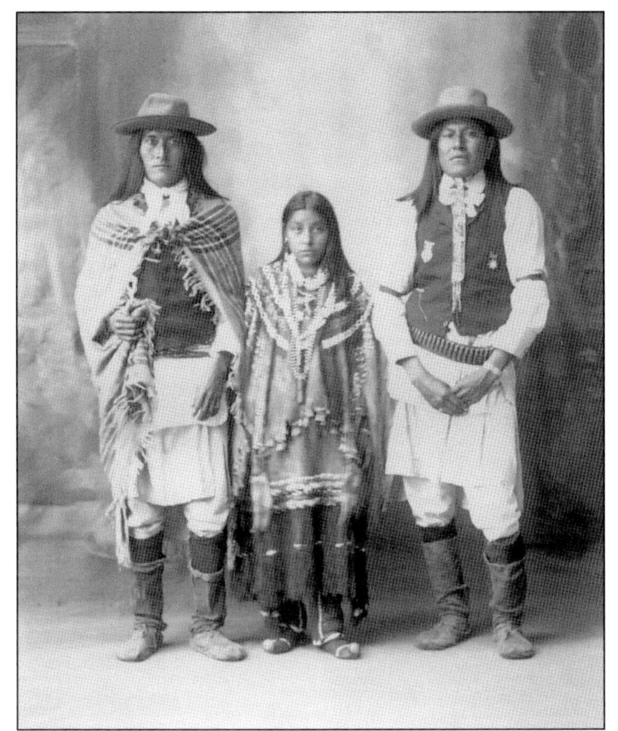

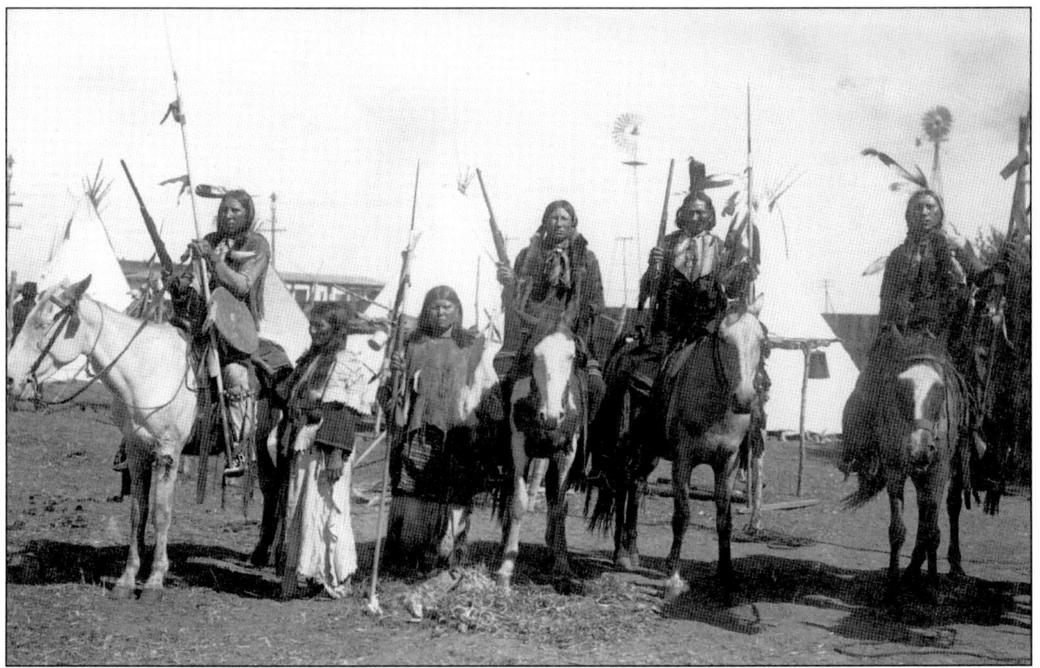

Twenty-two members of the Kiowa Apache tribe from southwest Oklahoma were in attendance at the Indian Congress. Visitors admired the buckskin clothing, which included detailed bead and fringe work, of the Kiowa Apache. Horses and ponies were also brought to the exposition, particularly for those tribes, such as the Kiowa Apache, who used animals during hunting expeditions. (Photo courtesy of Omaha Public Library.)

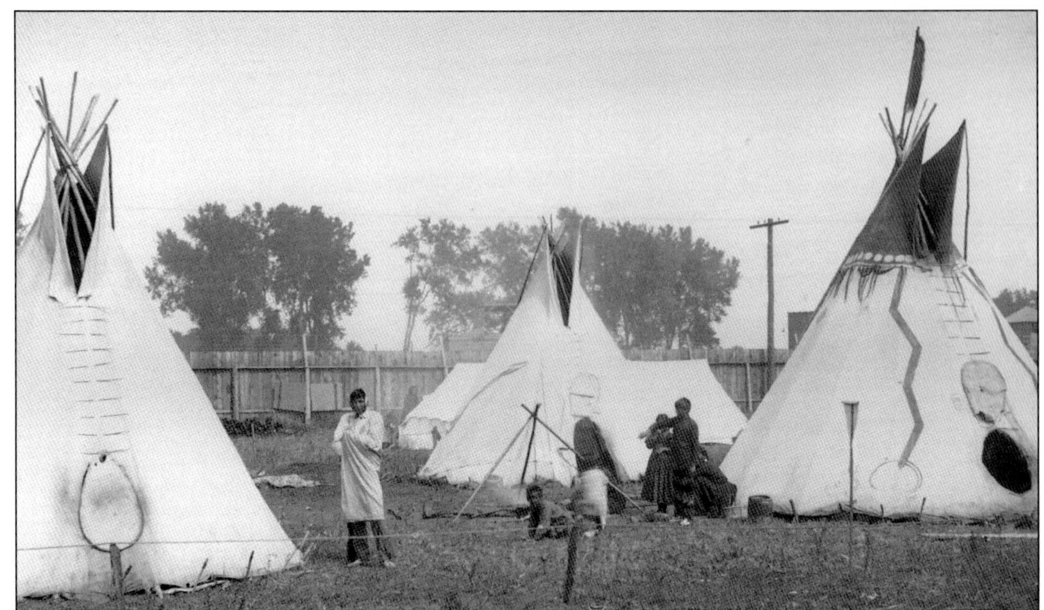

On the grounds of the Indian encampment, Assiniboine children play in the area between tepees. Amidst the Assiniboine camp, an example of a heraldic tepee was a principal attraction for both scholars and casual exposition visitors. (Photo courtesy of Omaha Public Library.)

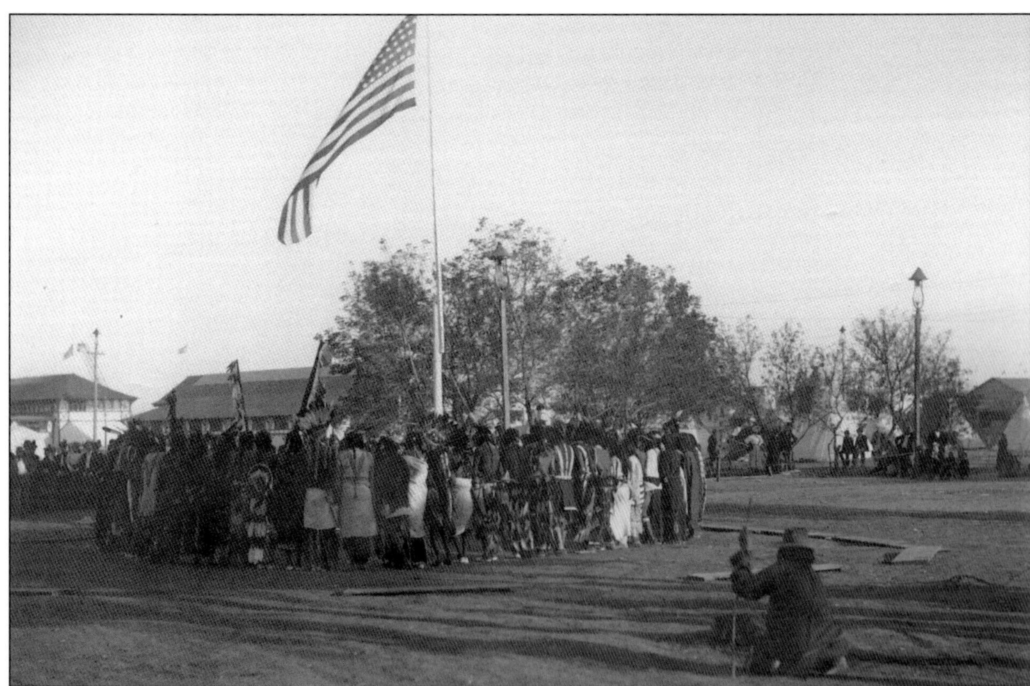

Surrounding a raised flag, Arapaho and Cheyenne representatives to the Indian Congress took part in the rituals of the Ghost Dance movement. Memories of the tragic events at Wounded Knee, eight years earlier, were still foremost in the minds of many Native Americans, who were experiencing the stresses associated with being relocated to reservations. (Photo courtesy of Omaha Public Library.)

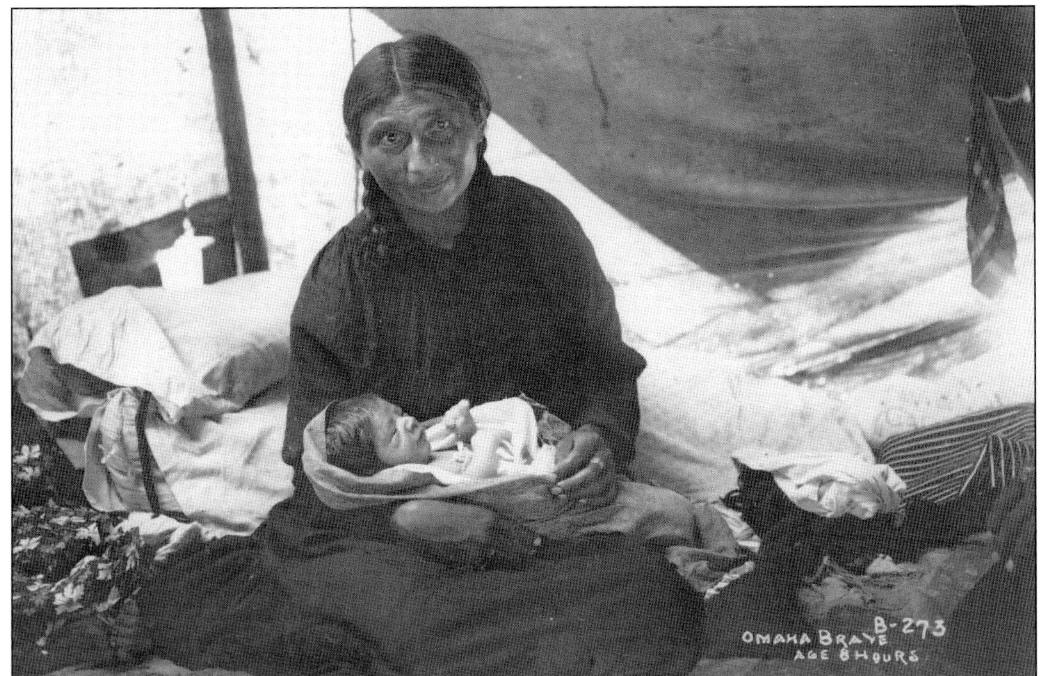

The spouse of Spotted Back cradles her son, eight hours hold, who held the distinction of being one of two infants born during the Indian Congress. Mother and son possessed an uncommon bond, as she was born less than a mile from the future site of the Indian Congress while traveling with the Omaha tribe 32 years earlier. (Photo courtesy of Omaha Public Library.)

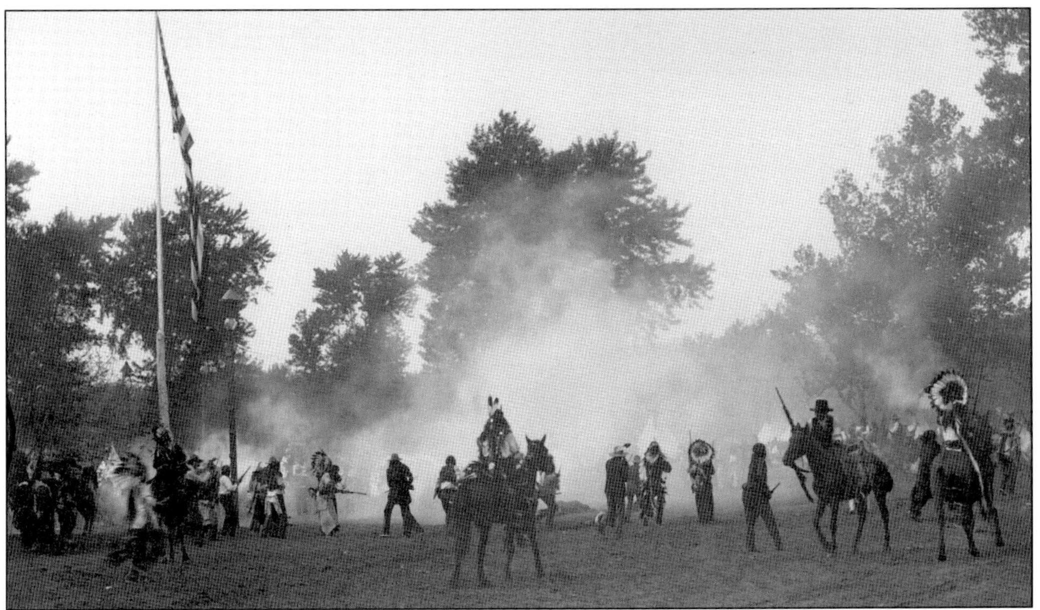

Sham battles were an exciting theatrical element of the Indian Congress. Members of various tribes participated in these demonstrations of equestrian skill and battle technique, amid a frenzy of blank rifle fire. One such battle was presented during the exposition visit of President McKinley for the occasion of Peace Jubilee Week. (Photo courtesy of Omaha Public Library.)

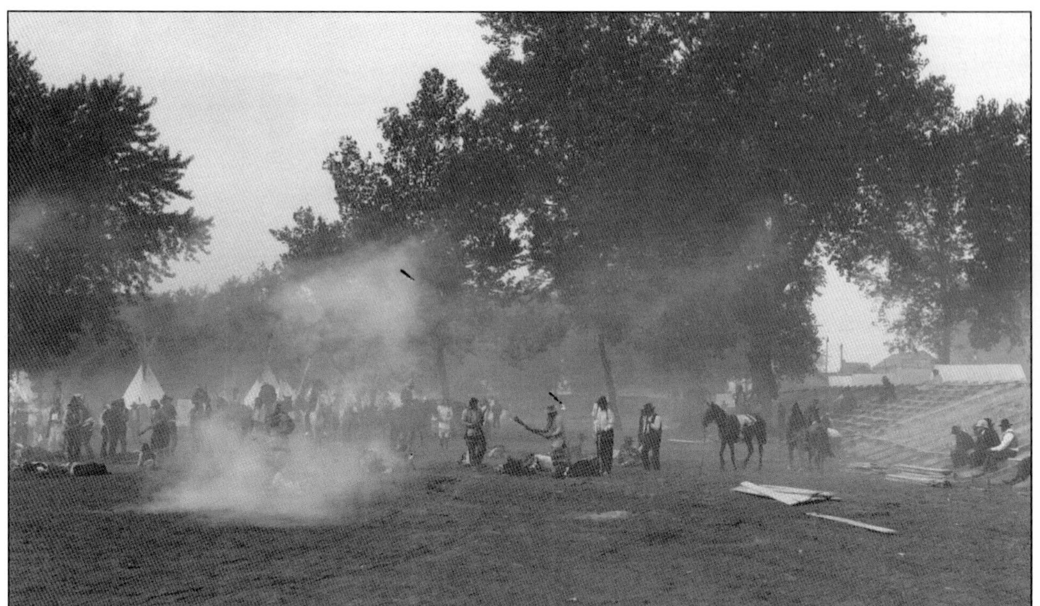

A small group of observers, pictured at the right, look on from bleacher-style seating as a Native American sham battle takes place. An element of suspense was often incorporated into the presentation of these sham battles when a prisoner was captured and later rescued by members of his tribe. (Photo courtesy of Omaha Public Library.)

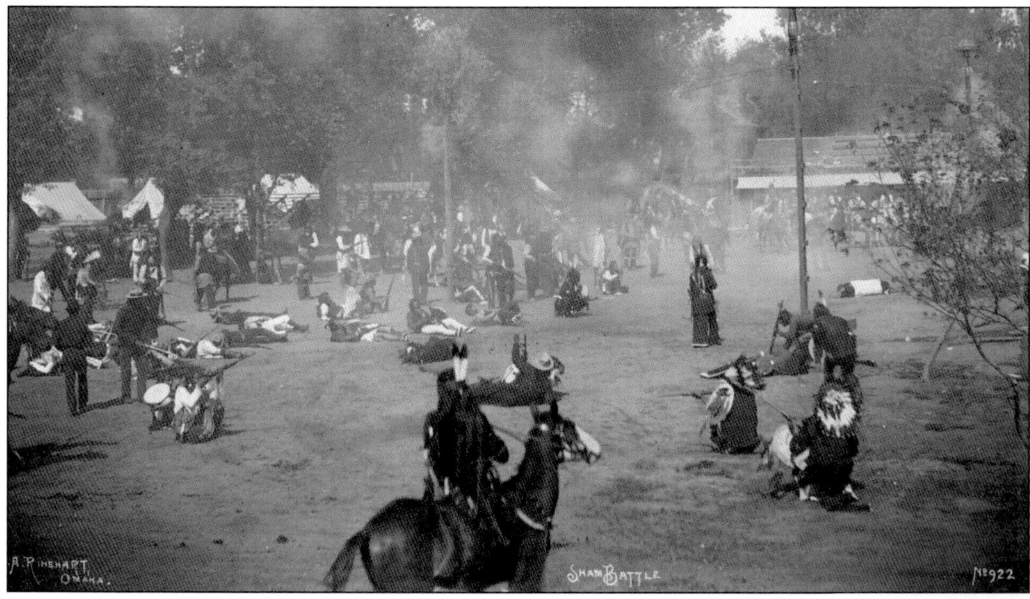

One Native American sham battle was described as a conflict between allied members of the Sioux, Wichita, Kiowa, and Assiniboine tribes against the Blackfoot. A prisoner of the Blackfoot was captured, tied to a stake and threatened. Blackfoot warriors, in an effort to rescue their tribesman, became involved in a battle with members of the opposing tribes. Though historical accuracy was not always the goal, many exposition visitors who witnessed a sham battle became consumed with what they regarded as the realistic nature of these performances. (Photo courtesy of Omaha Public Library.)

Six

THE WEST, REPRESENTED

Speaking at the dedication ceremony for the Nebraska Building, held early in the exposition season, Omahan William F. Gurley exclaimed, "Today, Nebraska throws open wide her golden gates and summons to her portals the myriads of mankind. To this enchanted city of the plains, she lures with wizard wand the unnumbered host of other lands and climes."

Inside of the nine state buildings positioned on the Bluffs tract were the offices of official delegations to the exposition, as well as a variety of displays open to public inspection. Displays representing 22 additional states were housed in the other significant buildings of the exposition, including those along the Grand Court.

The Horticulture Building, located immediately south of the state buildings, included many attractive floral and plant-life displays and was a temporary natural refuge from an exposition of advanced thought and seasonal celebration. The popular Transportation Building on the North tract of the exposition grounds featured exhibits in which visitors of all ages shared interest. While modern railroading was not unfamiliar to a city with close ties to the Union Pacific Railroad, such as Omaha, the opportunity to see locomotives and various types of railroad cars was eagerly anticipated by many exposition visitors. Approximately 520,000 visitors took the opportunity to walk through a deluxe passenger train during their visit to the Transportation Building.

Reduced railroad passenger rates attracted many visitors to Omaha for the Trans-Mississippi Exposition. Upon arrival, visitors were greeted with an exposition unlike any other event ever witnessed in the Great Plains.

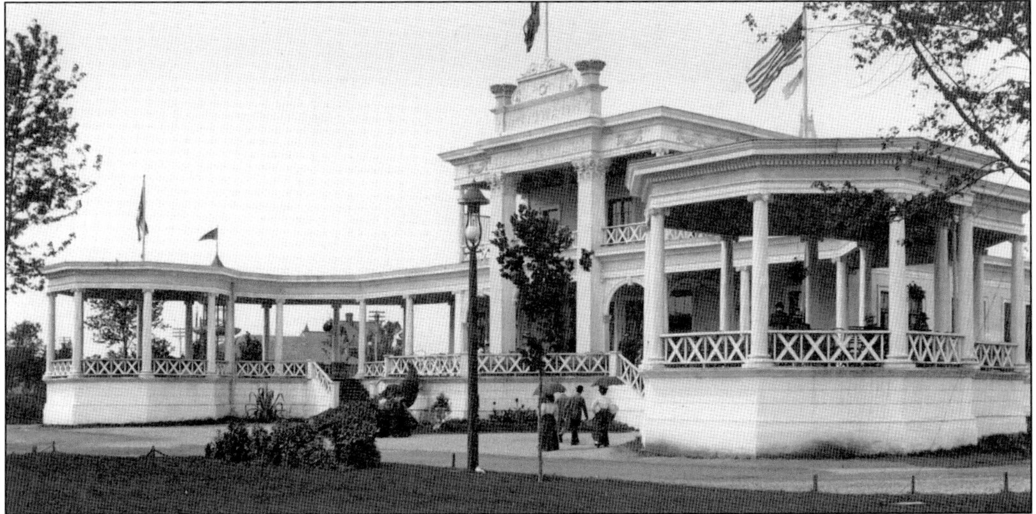

The 270-foot elliptical porch of the Iowa Building surrounded many attractive landscaping features, including extensive flower displays. Inside, visitors viewed private collections, displays of specific societies, and items representing the work of Iowa school children. (Photo courtesy of Omaha Public Library.)

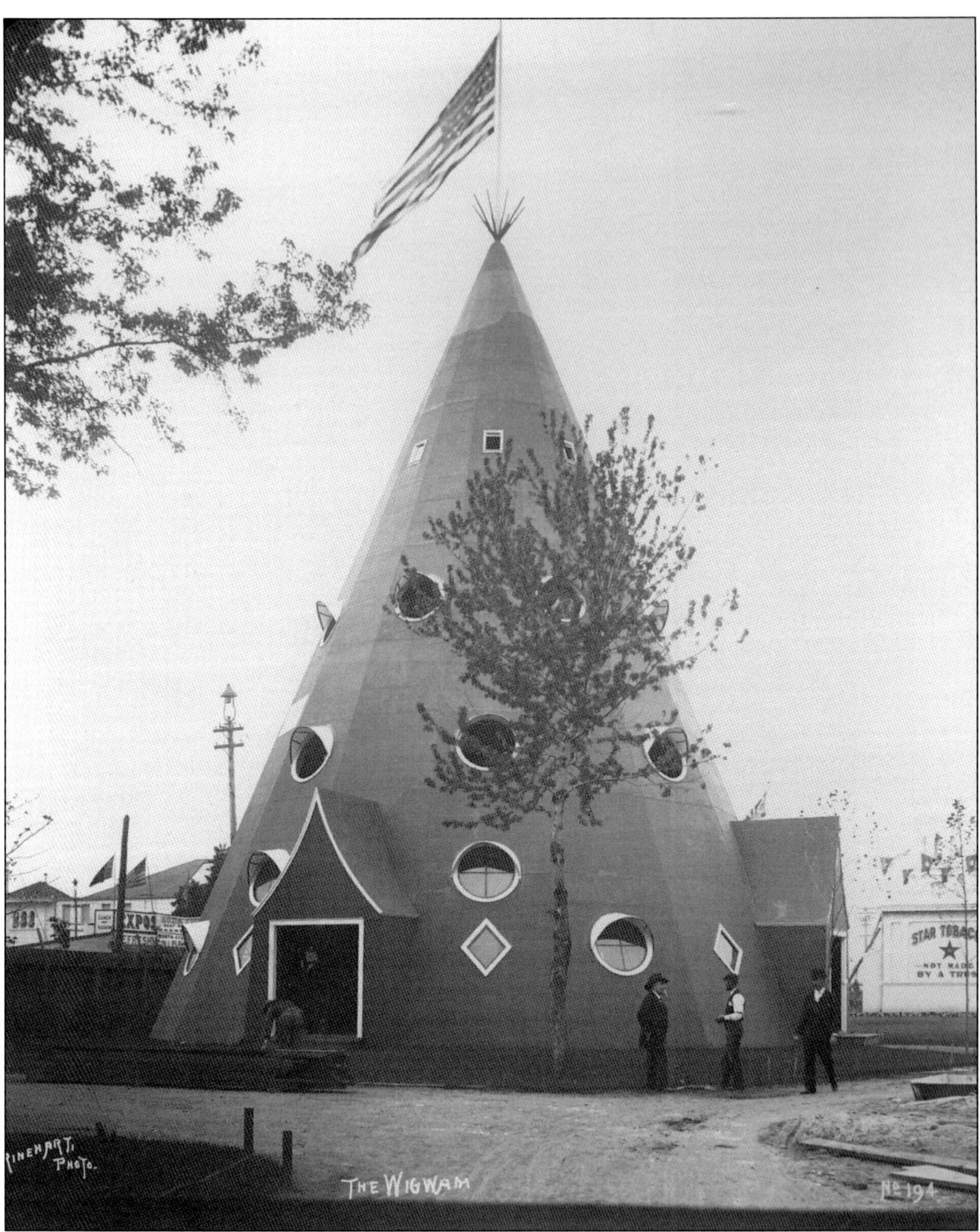

This 83-foot-high wigwam, representing Pottawattamie County, Iowa, was positioned on the exposition grounds near the Iowa Building. Constructed by the Exposition Association of Council Bluffs, Omaha's neighbor to the east, the structure contained four levels. The first two levels were devoted to displays, including a display of Native American culture, while the third and fourth floors provided for the comfort and relaxation of visitors. (Photo courtesy of Omaha Public Library.)

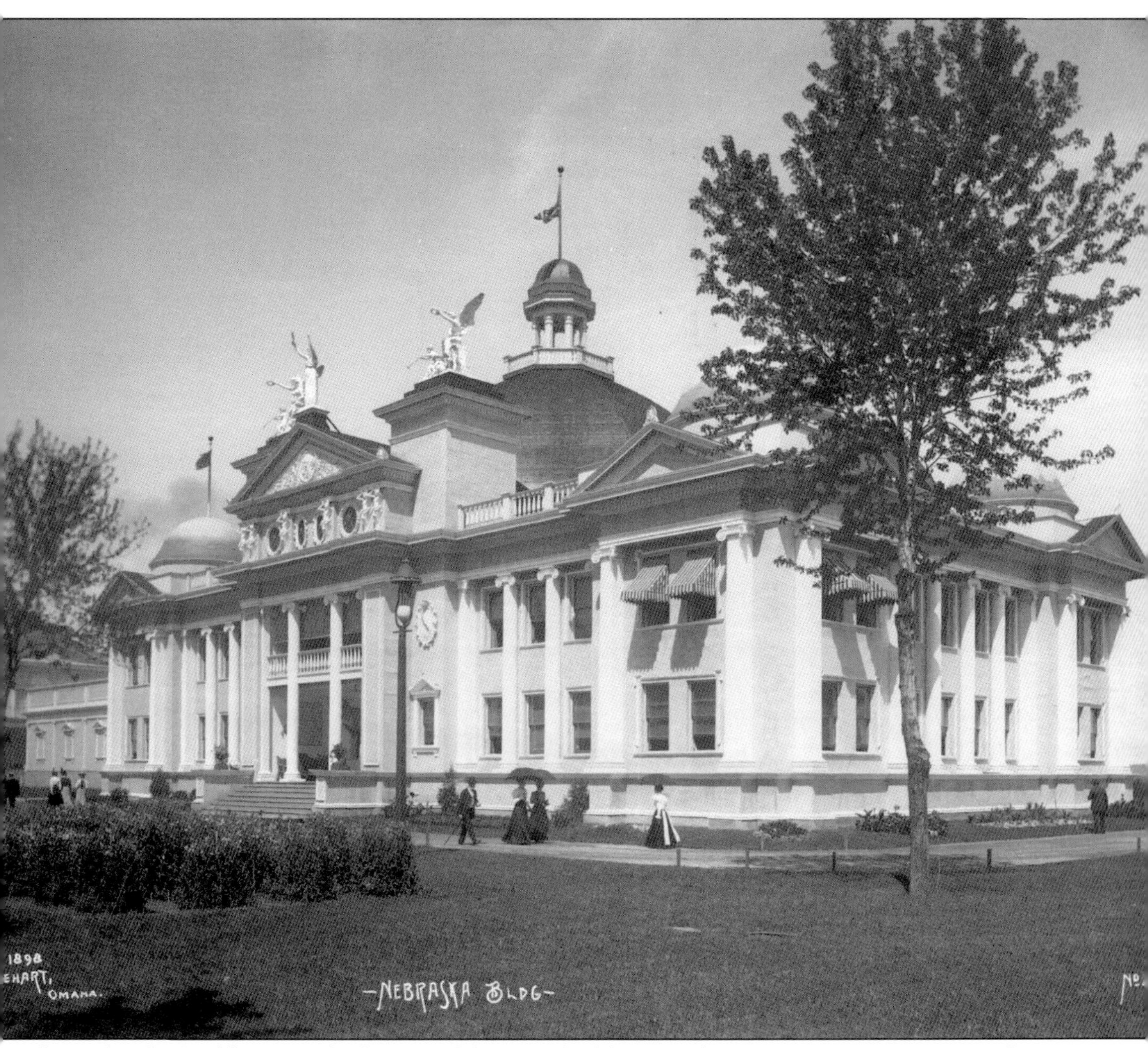

The Nebraska Building was the largest of the state buildings at the exposition. Measuring 90 feet by 145 feet, this structure included an octagonal dome with a diameter of 60 feet. Speaking at the building dedication ceremony on June 14, 1898, were Nebraska Governor Silas A. Holcomb and the former Democratic Presidential candidate from Nebraska, William Jennings Bryan. Exhibits by the Nebraska State Historical Society and the Daughters of the American Revolution were included in this primarily administrative building. (Photo courtesy of Omaha Public Library.)

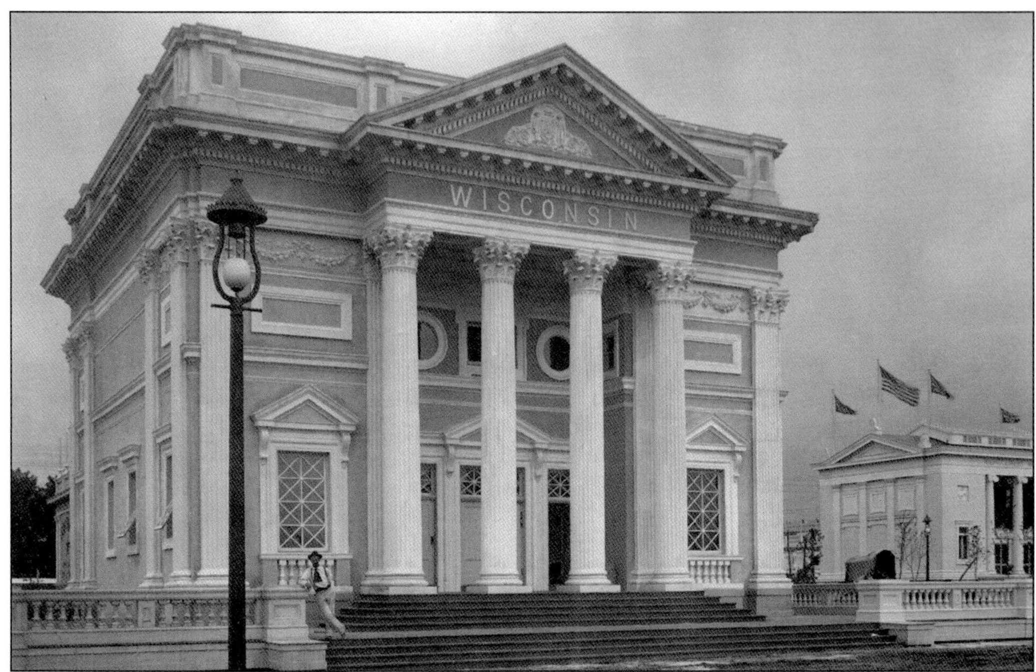

Described in the official exposition guide book as "another happy example of the enterprise of states not included in the original plan, which contemplated exhibits only from the commonwealths west of the Father of Waters," the classic design of the Wisconsin Building resulted from $15,000 in individual contributions. Displayed inside were private art collections and a fountain, lit beneath a central dome. (Photo courtesy of Omaha Public Library.)

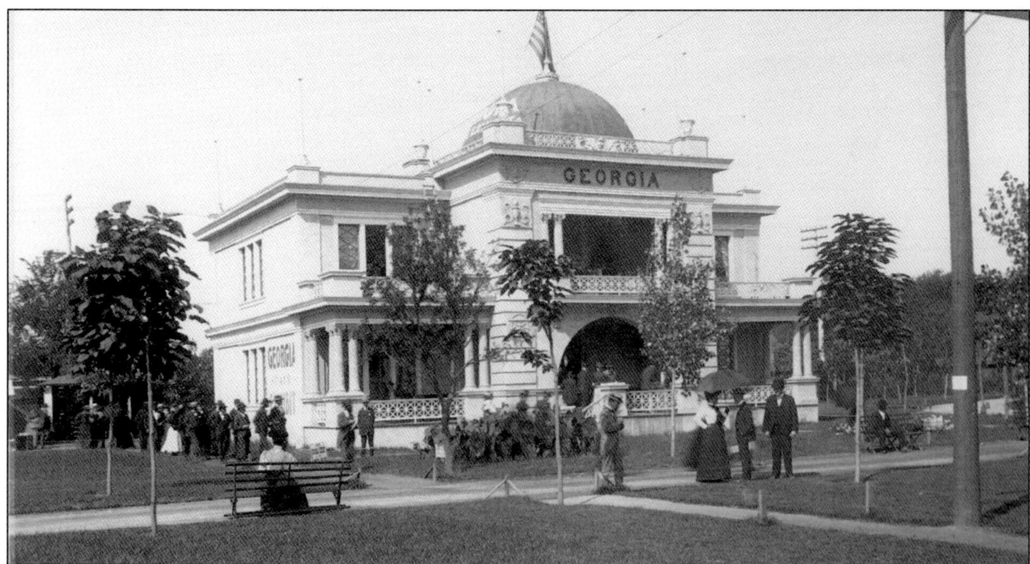

Constructed for the purpose of exhibition, the two floors of the Georgia Building showcased the varied industries of this state. The first floor promoted the lumber, mining, and stonework industries of Georgia, while the upper floor outlined the cotton production process from field to consumer. Additional agricultural displays on the second floor touted the achievements of Georgia's farmers. (Photo courtesy of Omaha Public Library.)

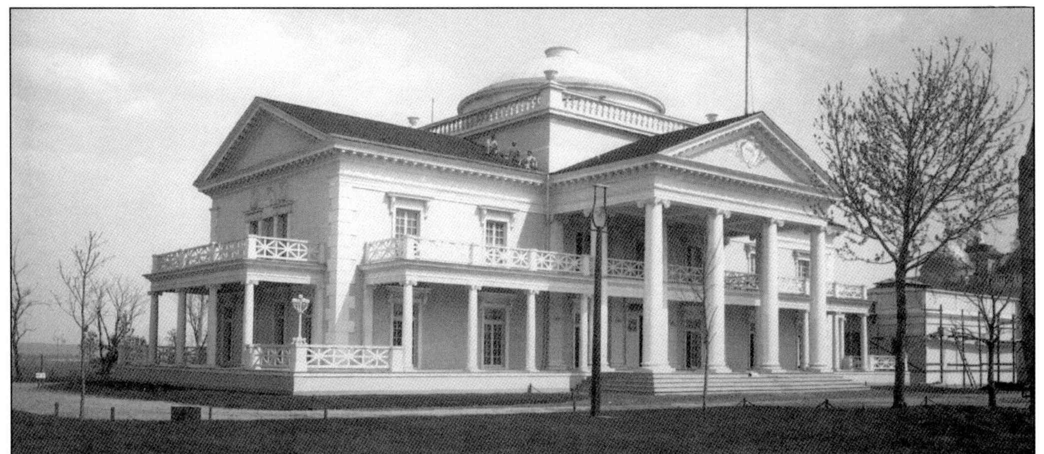

Dedicated on June 21, 1898, the Illinois Building was the product of one of the most significant single-state contributions to the exposition, amounting to $45,000. Located on the Bluffs tract, overlooking the Missouri River, this predominantly Colonial-style building included an annex which housed a group of pictures painted by John R. Key on the occasion of the 1893 Chicago World's Fair. (Photo courtesy of Omaha Public Library.)

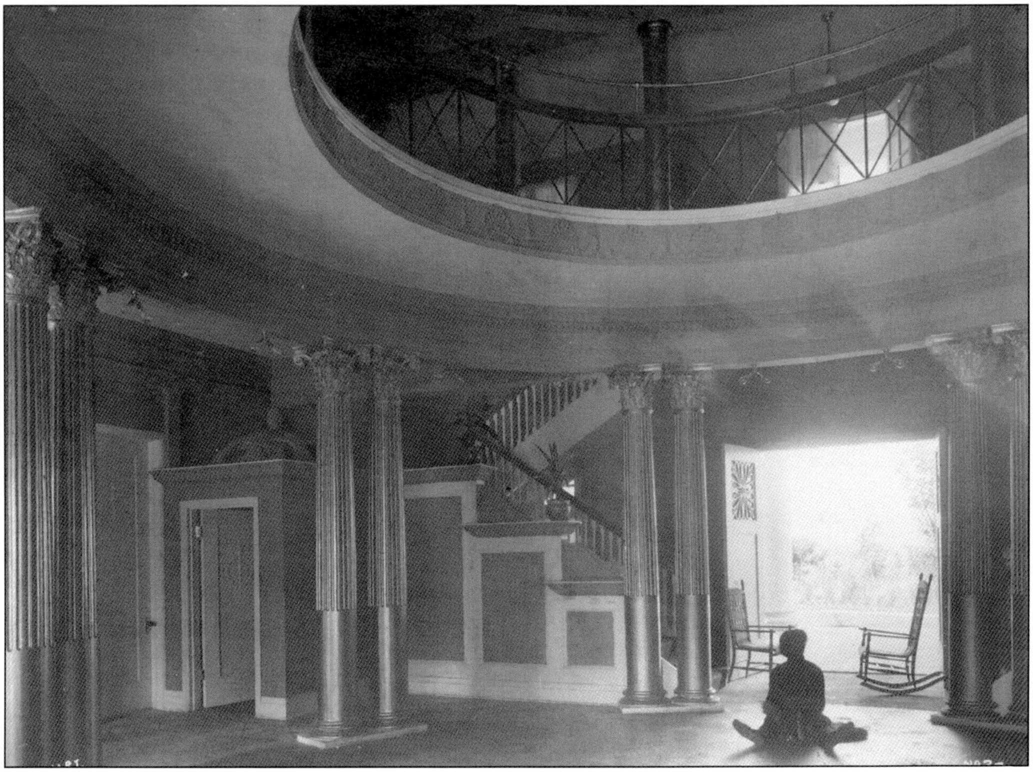

The interior of the Illinois Building included a central rotunda that served as a reception room for visitors. Upon dedication, fine furniture, draperies, and other notable design elements added to the welcome experienced by visitors from Illinois and throughout the world. The first floor of this structure included parlors and the private offices of Illinois exposition officials. The second floor featured private meeting rooms and additional offices. (Photo courtesy of Omaha Public Library.)

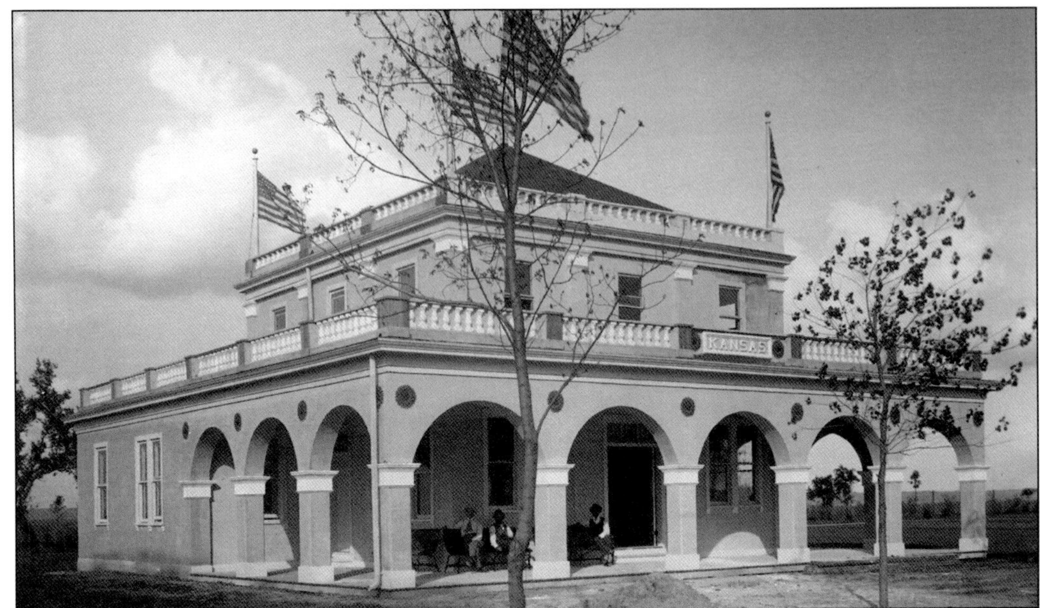

Resting in the shade on the patio of the Kansas Building, these gentlemen must have enjoyed the opportunity to rest and appreciate the view of the nearby Horticulture Building. The two-story state building measured 55 feet by 57 feet. This California-mission-style structure included a photographic display of Kansas farm life, as well as paintings depicting the commercial achievements of this state. (Photo courtesy of Omaha Public Library.)

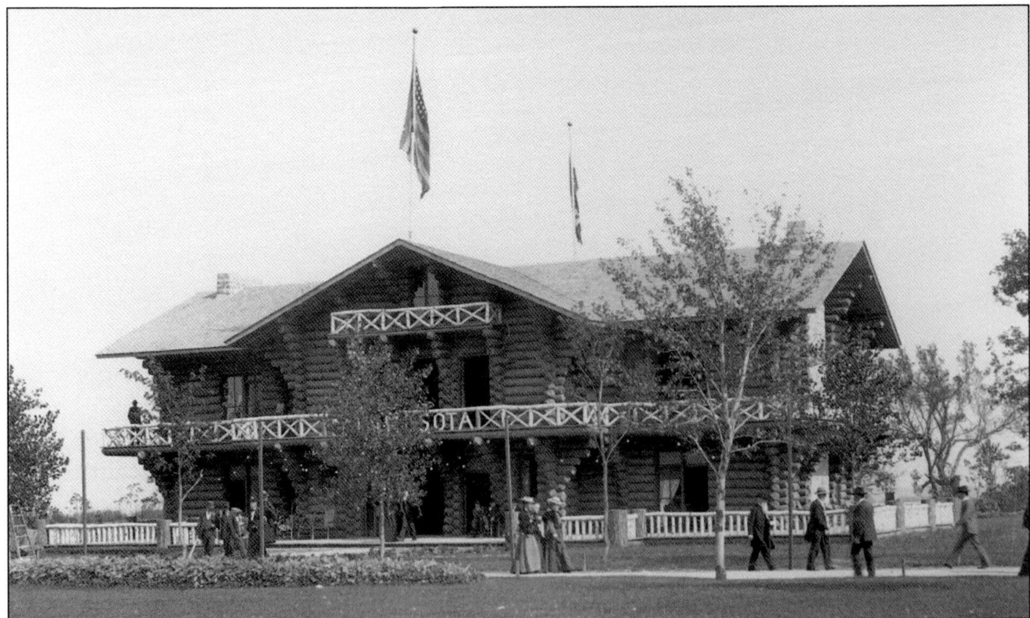

The unique log cabin representing the state of Minnesota was constructed of logs measuring 12 to 14 inches in diameter that remained in their natural condition. The two-story structure measured 60 feet by 70 feet and was designed to represent life in the forests. Appropriate reception space and parlors on the first floor welcomed visitors to displays of forest life featuring the materials necessary for logging, hunting, and cooking in this setting. (Photo courtesy of Omaha Public Library.)

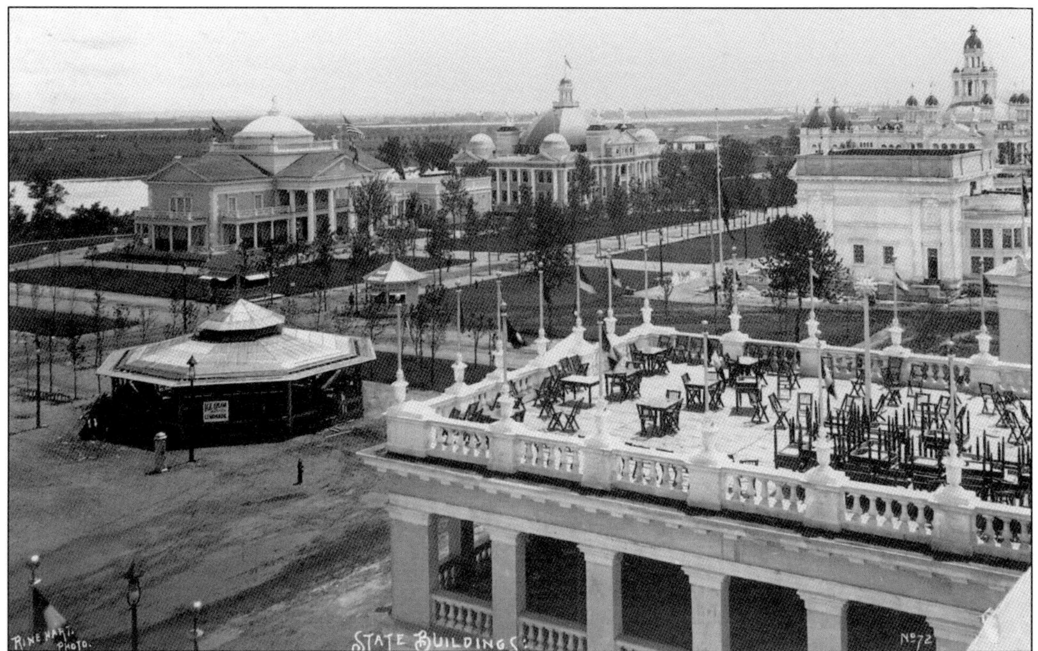

While 31 states were represented with displays at the exposition, nine states (Georgia, Iowa, Illinois, Kansas, Minnesota, Montana, Nebraska, New York, and Wisconsin) were additionally represented by individual buildings. This area at the east side of the exposition grounds was known as the Bluffs tract and was the location of all nine state buildings. Pictured at the lower right is rooftop seating of the three-story restaurant at the South Viaduct. (Photo courtesy of Omaha Public Library.)

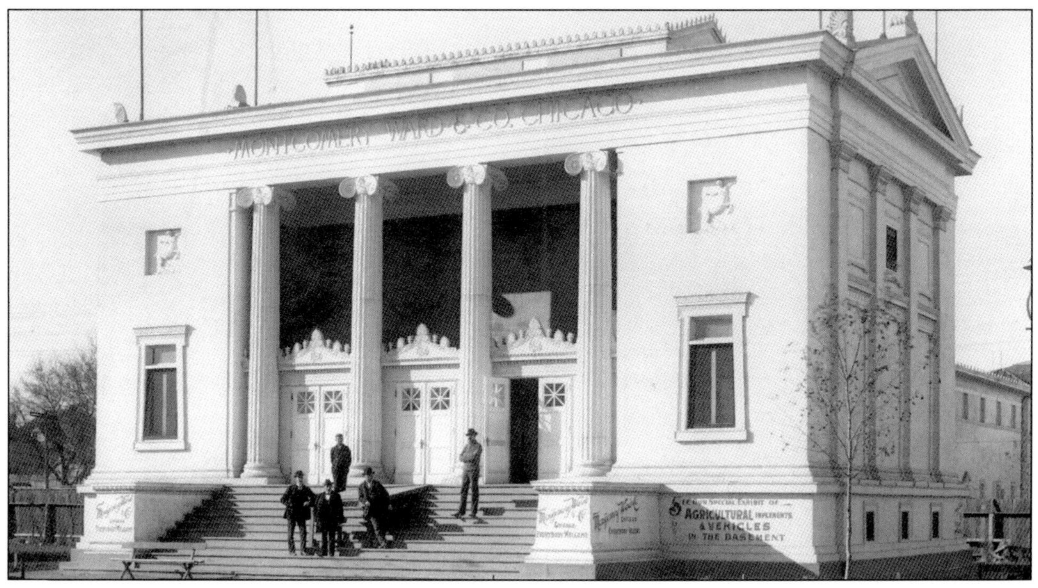

The Montgomery Ward and Company Building, located along the highly visible area of the Grand Plaza, featured the products of this large retailer. An extensive display of agricultural tools, including vehicles, was featured on the lower floor of this structure. (Photo courtesy of Omaha Public Library.)

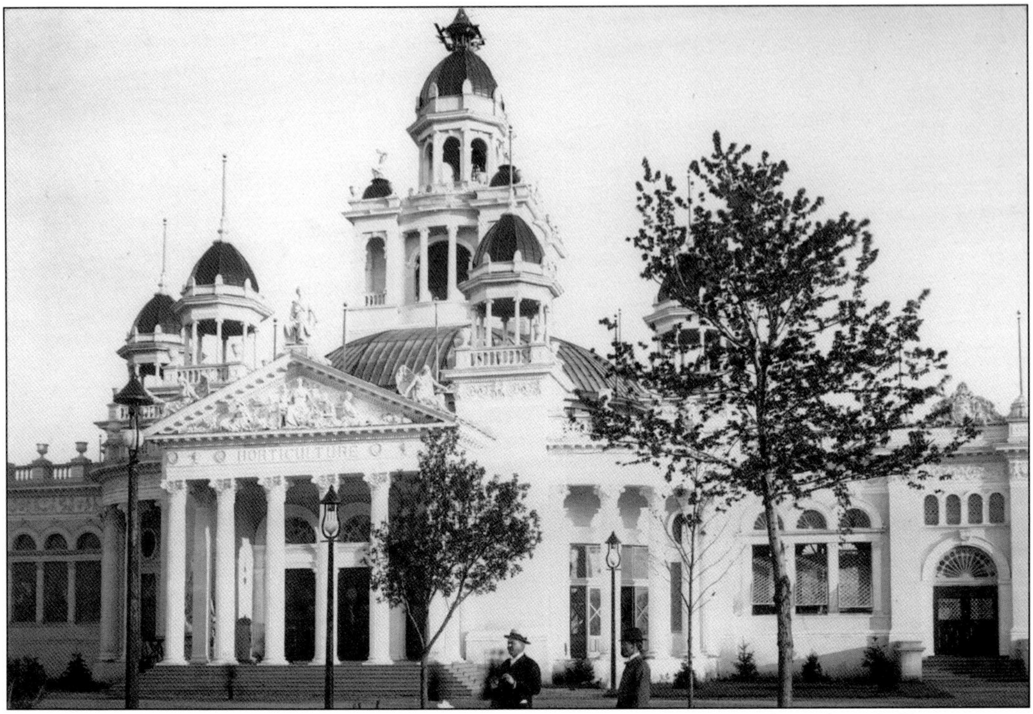

The Horticulture Building, measuring 310 feet by 130 feet and situated on the Bluffs tract, was surrounded by state buildings. At the top of the central dome was a belfry, the chimes of which rang out across the exposition grounds from a height of 160 feet. Skylights provided natural illumination for the plants beneath the dome in this structure. The balcony immediately above the dome featured an observatory. (Photo courtesy of Omaha Public Library.)

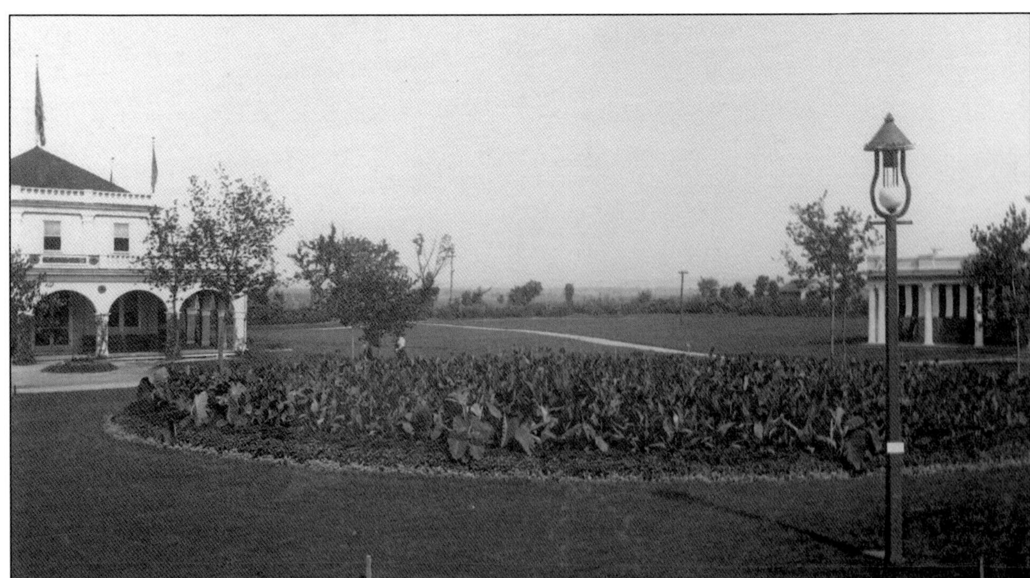

This foliage display was on the east side of the Horticulture Building, in the open space between the Kansas Building (left) and the New York Building (right). (Photo courtesy of Omaha Public Library.)

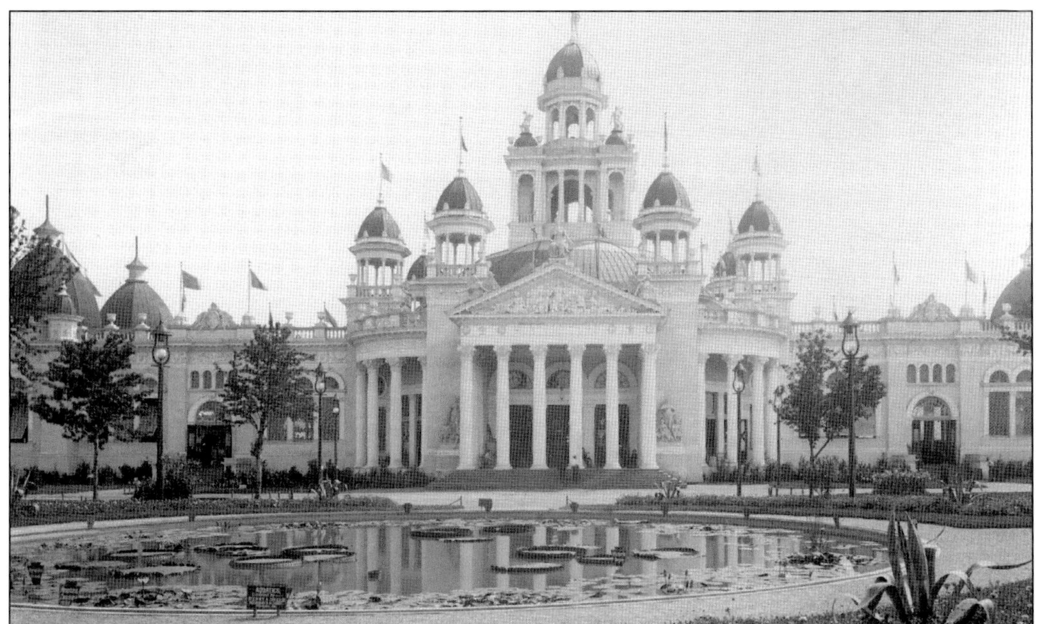

The gold medal-winning aquatics exhibit of Philadelphian Henry A. Dreer was the focal point at the northern entrance of the Horticulture Building. Eight minarets, four of which are pictured here on either side of the main entrance, surrounded the central dome. Ornamentation elements incorporated into the exterior design of this structure celebrated the fruits, flowers, and plants on display in its interior. (Photo courtesy of Omaha Public Library.)

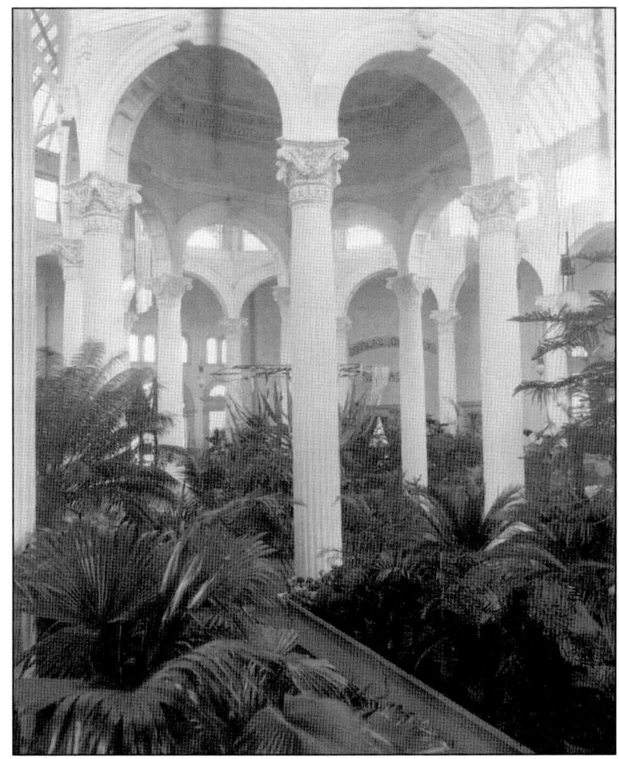

Displays inside the Horticulture Building changed frequently, based upon the availability of plants and the need to preserve the floral appearance of exhibits. Six hundred exhibitors displayed native and semi-tropical plants, as well as the harvested fruit of represented regions. (Photo courtesy of Omaha Public Library.)

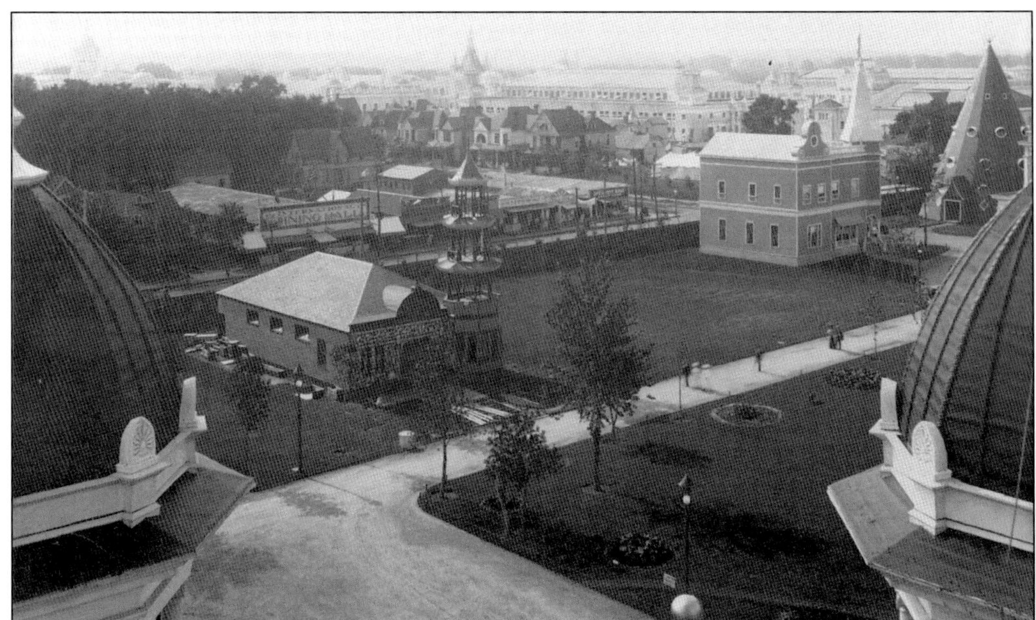

The observatory of the Horticulture Building provided this spectacular view of the exposition grounds and surrounding districts. The single-story structure with a tower, pictured at the left, offered Chinese souvenirs of ivory and bamboo furniture. The two-story, wood structure with a light-colored tower, pictured at the right, was the Montana Building, featuring office space decorated with paintings and hunting trophies from the collections of state mine owners. (Photo courtesy of Omaha Public Library.)

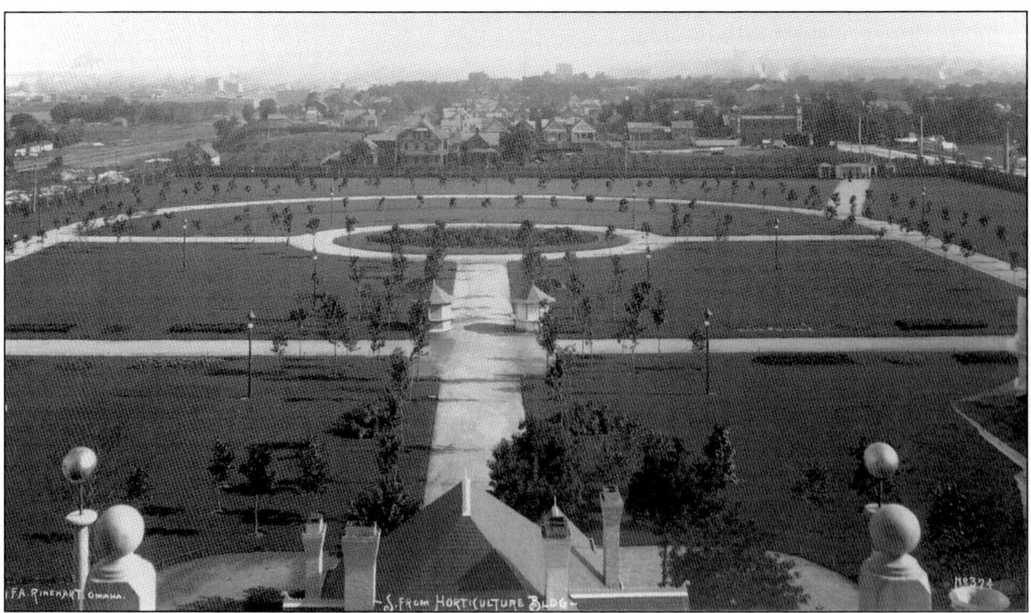

This southerly view toward downtown Omaha, as seen from the observatory of the Horticulture Building, depicts the efforts of exposition designers to incorporate open space and walking paths into the design of the exposition grounds. Homes along Binney Street are pictured in the distance. (Photo courtesy of Omaha Public Library.)

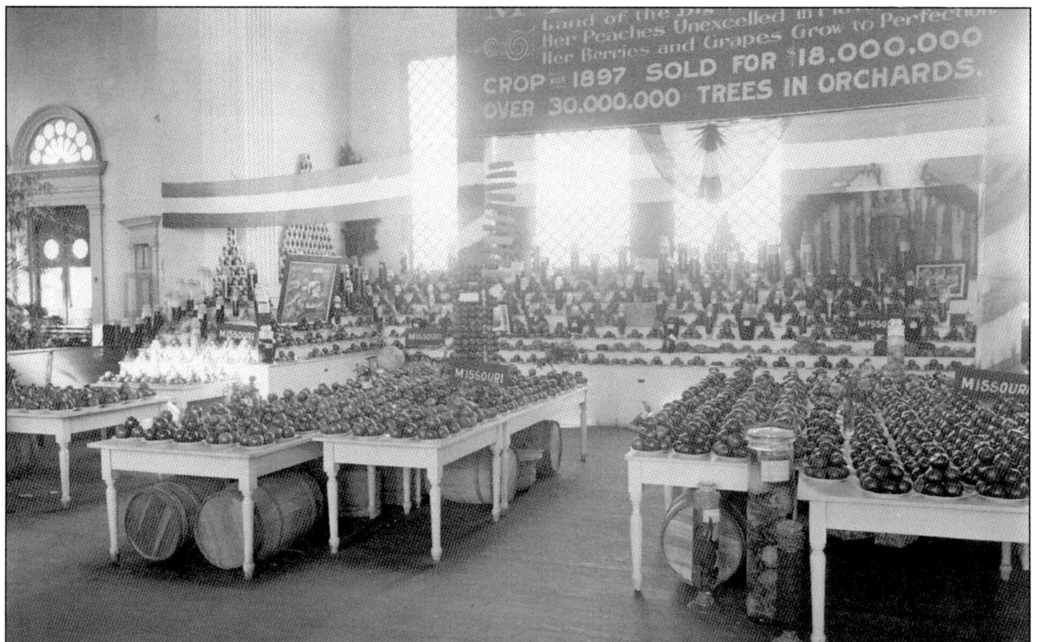

Thirty varieties of apples, cherries, peaches, plums, and assorted berries were displayed in the horticultural exhibit representing the state of Missouri. The banner above the produce display boasts the yields of an 1897 crop, valued at $18 million from 30 million trees. (Photo courtesy of Omaha Public Library.)

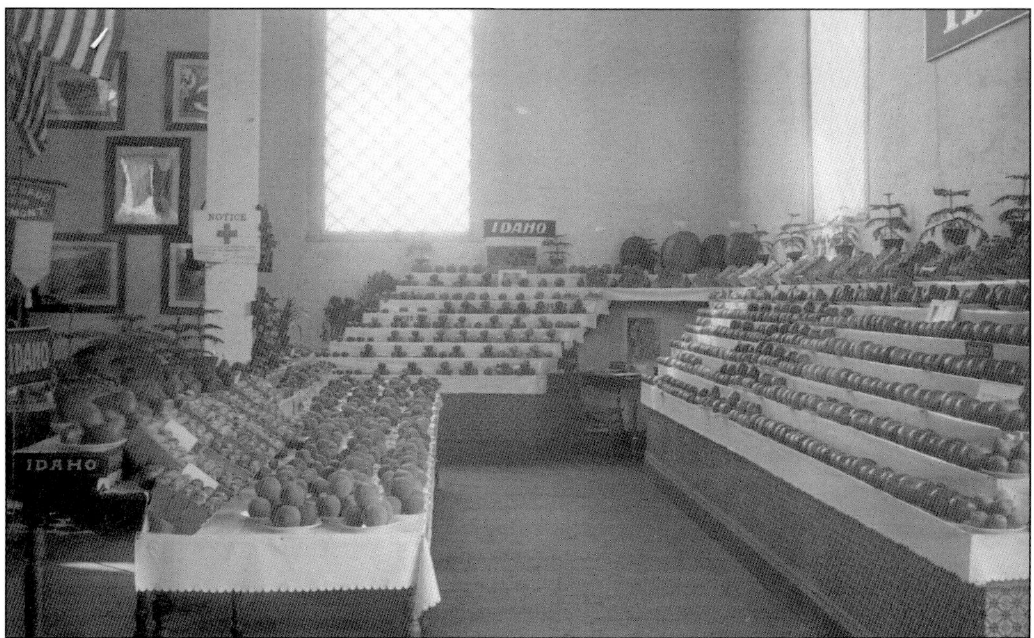

Available for inspection to visitors of the Horticulture Building were apples, peaches, pears, and pumpkins from the state of Idaho. Fresh products were periodically shipped to exhibitors to maintain freshness and present the products of the state at their best. (Photo courtesy of Omaha Public Library.)

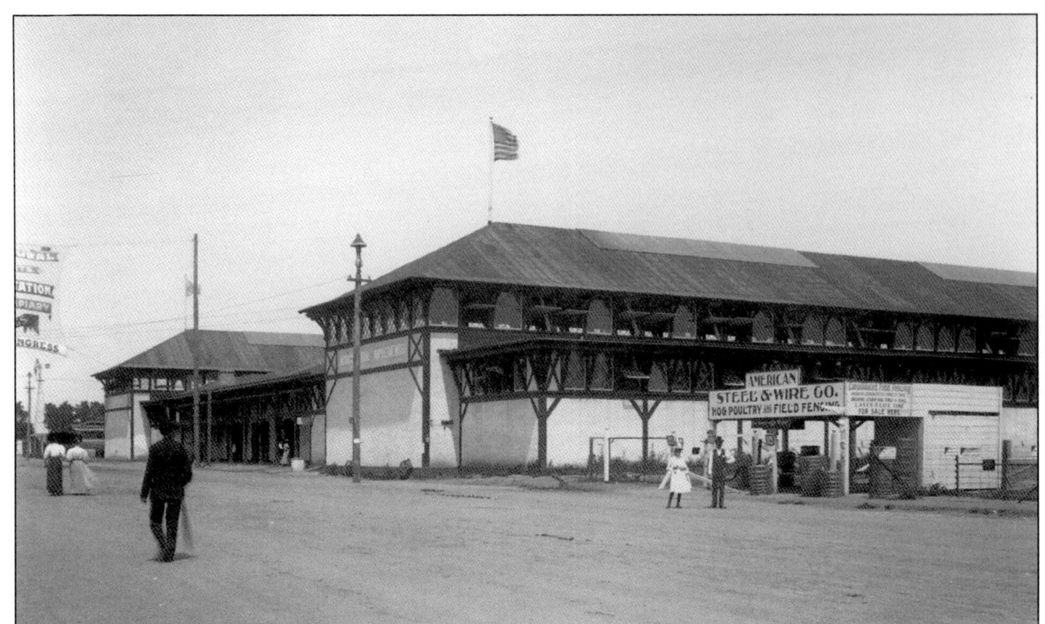

The half-timbered Transportation and Agricultural Implement Building, the largest of the exposition buildings, combined plaster with its wooden frame and featured no decorative elements similar to the structures of the Grand Court. Vehicles, farm machinery, locomotives, and other railway cars were featured in this building that included four railroad entrances and 1,728 feet of track. (Photo courtesy of Omaha Public Library.)

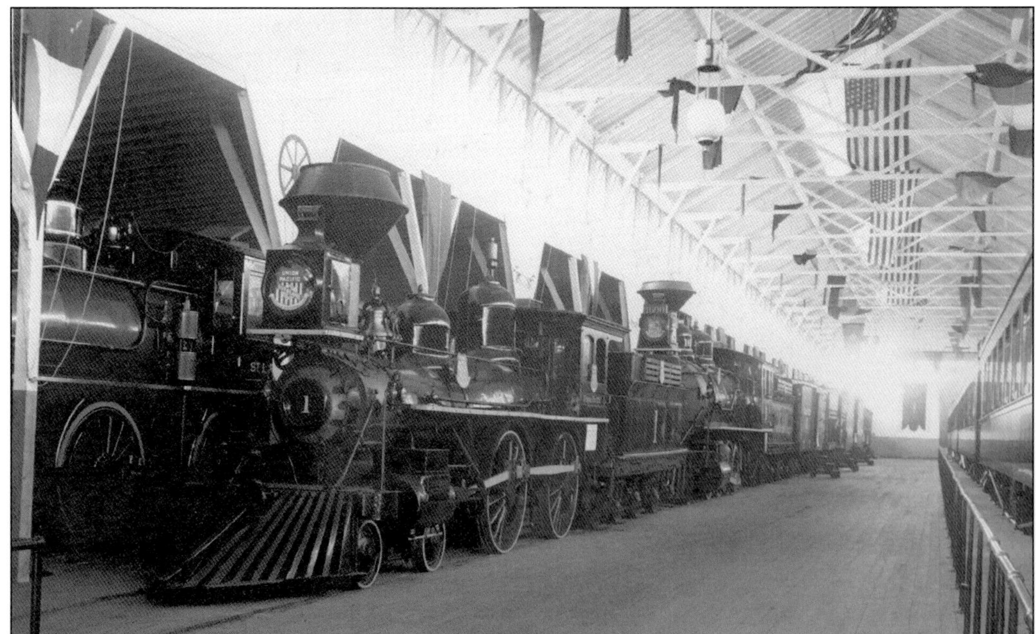

This Union Pacific locomotive and accompanying cars were a source of pride for local visitors who, thirty years previous, participated in the construction of the first transcontinental railroad after Omaha was selected as its eastern terminus. The opportunity to walk through and see the trains up close was an attraction for many visitors, young and old. (Photo courtesy of Omaha Public Library.)

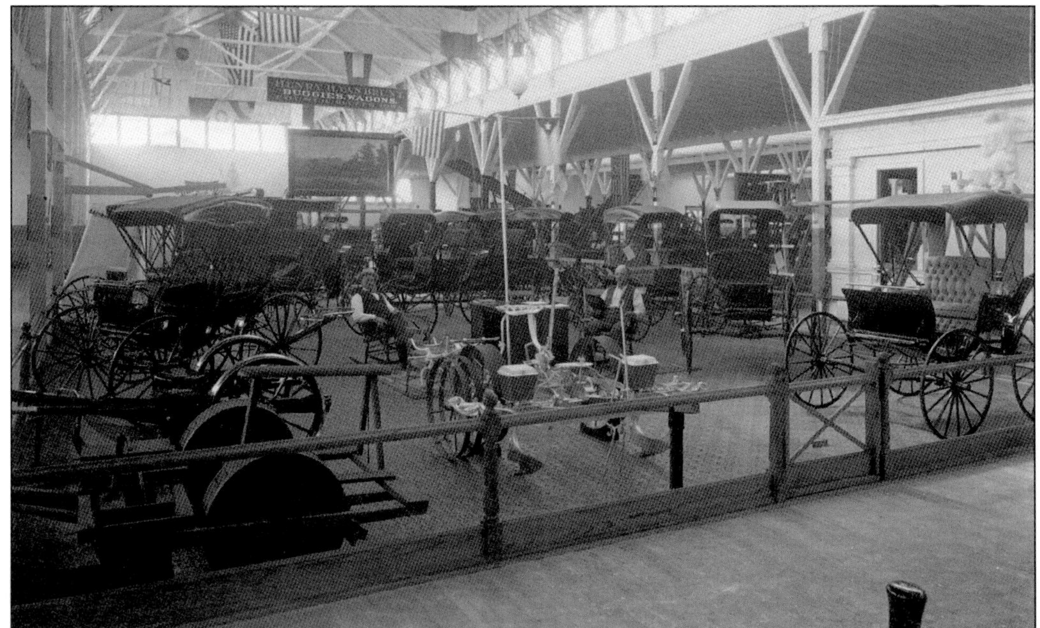

Light-colored fabric, in place of glass windows, was used to illuminate the center section of the Transportation and Agricultural Implement Building. This display of buggies and wagons by Henry H. Van Brunt of Council Bluffs, Iowa, attracted visitors who were interested in the most modern means of transportation. (Photo courtesy of Omaha Public Library.)

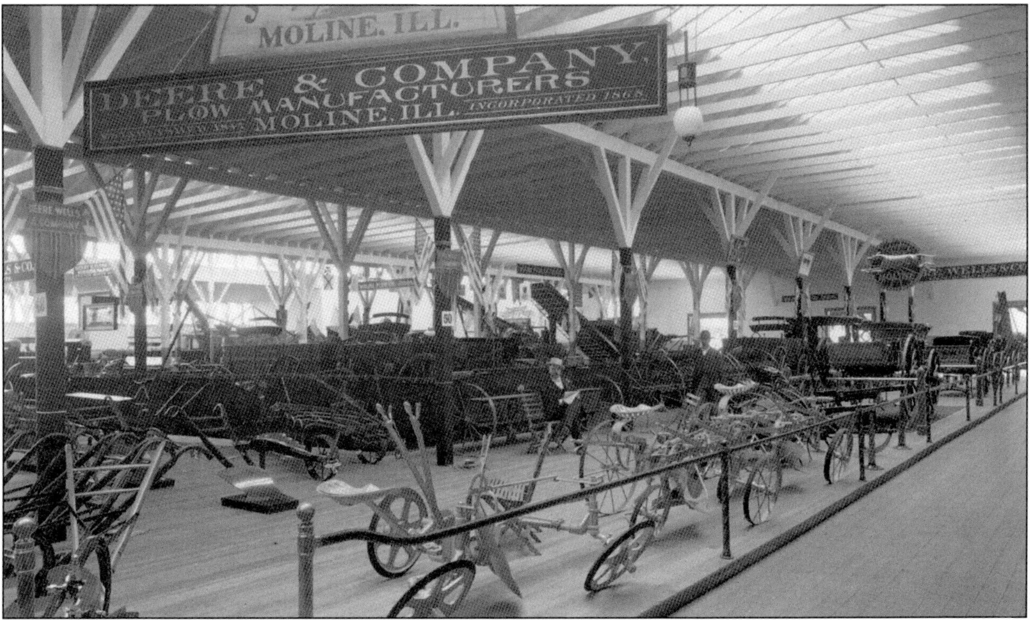

The Transportation and Agricultural Implement Building featured 151 exhibitors, who utilized 59,158 square feet of exhibition space. Included among displays of farm equipment were a modern Pullman vestibule train and the "Lincoln coach," an armor-clad coach used by President Lincoln during the Civil War and, later, used to transport his remains to Springfield, Illinois, for burial. (Photo courtesy of Omaha Public Library.)

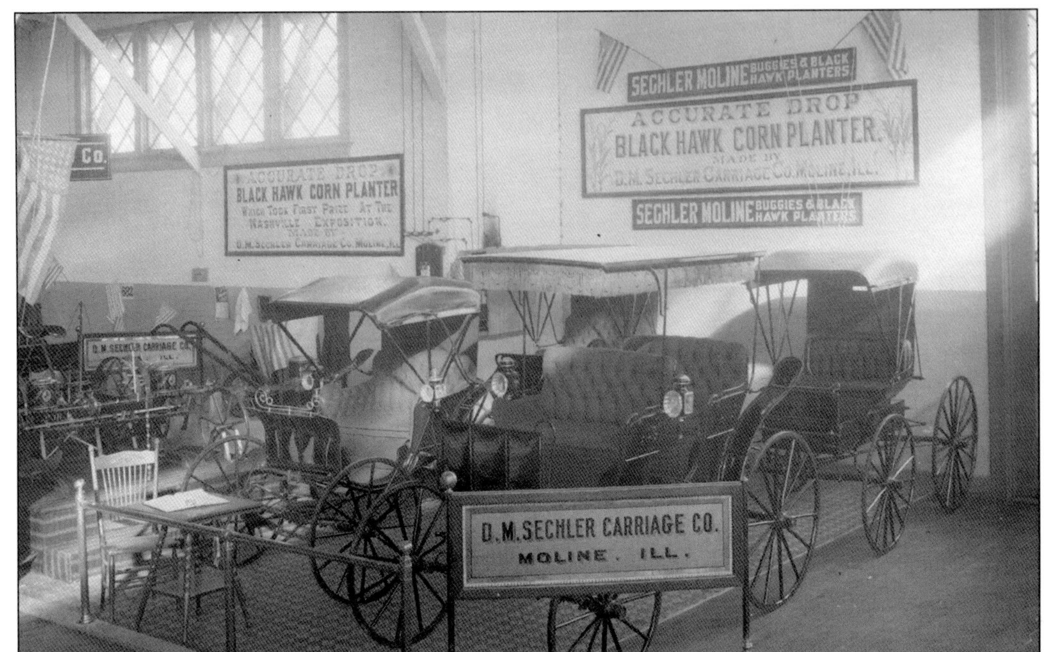

The D.M. Sechler Company of Moline, Illinois, displayed this collection of one- and two-seat carriages along with planting equipment inside of the Transportation and Agricultural Implement Building. (Photo courtesy of Omaha Public Library.)

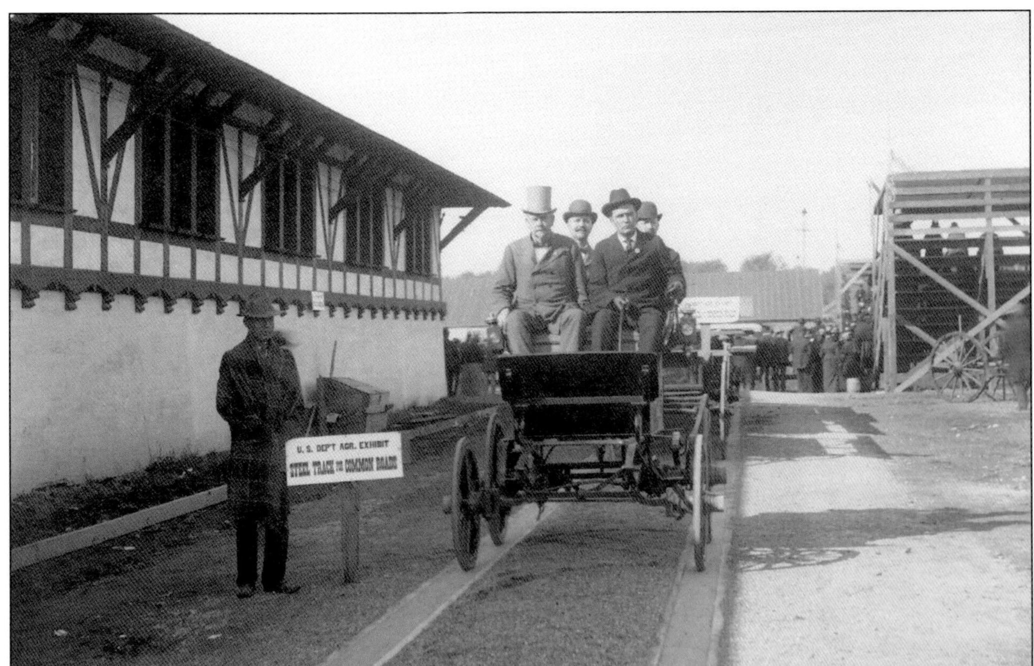

These four gentlemen enjoyed a ride in a horseless carriage along the steel tracks placed by the United States Department of Agriculture as part of an exhibit of modern roadway. (Photo courtesy of Omaha Public Library.)